D1000831

CHEMISTRY VERSUS PHYSICS

Chemical Reactions Near Critical Points

CHEMISTRY VERSUS PHYSICS

Chemical Reactions Near Critical Points

Moshe Gitterman

Bar-Ilan University, Israel

World Scientific

NEW JERSEY • LONDON • SINGAPORE • BEIJING • SHANGHAI • HONG KONG • TAIPEI • CHENNAI

Published by

World Scientific Publishing Co. Pte. Ltd.

5 Toh Tuck Link, Singapore 596224

USA office: 27 Warren Street, Suite 401-402, Hackensack, NJ 07601

UK office: 57 Shelton Street, Covent Garden, London WC2H 9HE

British Library Cataloguing-in-Publication Data
A catalogue record for this book is available from the British Library.

CHEMISTRY VERSUS PHYSICS
Chemical Reactions Near Critical Points

ISBN-13 978-981-4291-20-0
ISBN-10 981-4291-20-X

Printed in Singapore by B & Jo Enterprise Pte Ltd

Preface

In this book we attempt to trace the connection between chemical reactions and the physical forces of interaction manifested in critical phenomena. The physical and chemical descriptions of matter are intimately related. In fact, the division of forces into "physical" and "chemical" is arbitrary. It is convenient [1] to distinguish between strong attractive (chemical) forces leading to the formation of chemical species, and weak attractive (physical) forces, called van der Waals forces. It should be remembered, therefore, when one considers the "ideal" ternary mixture A, B and $A_m B_n$, that the strong chemical bonding interaction between A and B atoms has already been taken into account via the formation of the chemical complex $A_m B_n$, and the term "ideal" only means that there are no "physical" forces present. The growth of clusters in a metastable state is an example of the fuzzy distinction between physical and chemical forces. In numerical simulations a rather arbitrary decision has to be made whether a given particle belongs to a "chemical" cluster.

The usefulness of a "chemical" approach to physical problems can be seen from the mean-field theory of the phase transition on an Ising lattice of non-stoichiometric AB alloys [2]. The temperature dependence of the long-range and short-range order parameters is found from the "law of mass action" for the appropriately chosen "chemical reaction." The latter is the exchange of position of an atom A from one sublattice and an atom B from the second sublattice. The change of the interaction energy for such a transition when both atoms are or are not nearest neighbors determines the "constant of chemical reaction." Such an approach allows one to avoid the calculation of entropy provided that one is interested only in the value of the critical temperature, rather than in the behavior of all thermodynamic quantities, which are determined by the same classical critical indices in all versions of mean-field theory. The latter example is, in fact, a special case of chapter 15 in [3], where the typical "physical" process of diffusion is considered as a "chemical reaction" in which some amount of substance A passes from volume element a to b while a different amount of B passes from b to a.

The "chemical" method, in which a given atom with all its neighbors is considered as the basic group, gives better results than the Bragg-Williams or Bethe-Peierls method. In the Bragg-Williams method, each atom is exposed to the (self-consistent) average influence of all other atoms, whereas in the Bethe-Peierls method, a pair of adjacent atoms is considered as the basic group.

Even though the border between chemical and physical forces is arbitrary, one usually considers first the "physical" forces in the equation of state, and then the "chemical" forces in the law of mass action based on the non-ideal equation of state. I shall follow this approach.

One can also trace the common features of phase transitions and chemical reactions by analyzing the time evolution of the state variables ψ described by the equation

$$d\psi/dt = F\left[\psi, \lambda\right], \tag{0.1}$$

where λ is the set of internal and external parameters, and F is a nonlinear functional in ψ. In the case of a phase transition, Eq. (0.1) may be the Landau-Ginzburg equation for the order parameter, while in the case of a chemical reaction — the equation for the rate of reaction. The steady state of the system is described by $\psi^0(\lambda)$ which is the solution of equation $F\left[\psi^0, \lambda\right] = 0$. For nonlinear F, more than one steady state solution is possible, and for some values of λ, at the so-called transition point, bifurcation may occur, when a system goes from the original steady state to the new steady state. Such a transition might be of first or second order in the case of a phase transition, and, analogously, the hard or soft transition for a chemical reaction.

I have kept this book as simple as possible, so that it will be useful for a wide range of researchers, both physicists and chemists, as well as teachers and students. No preliminary knowledge is assumed, other than undergraduate courses in general physics and chemistry. In line with this approach, I have favored a phenomenological presentation, thus avoiding the details of both microscopic and numerical approaches. There are a tremendous number of published articles devoted to this subject, and it proved impossible to include many of them in this book of small size. I ask the forgiveness of the authors whose publications were beyond the scope of this book.

The organization of the book is as follows.

After the Introduction, Chapter 1 contains a short review of phase transitions and chemical reactions and their interconnection, which is needed to understand the ensuing material. Chapter 2 is devoted to the specific changes in a chemical reaction occurring near the critical point. The effect of pressure and phase transformation on the equilibrium constant and rate of reaction is the subject of Secs. 2.1–2.2. The hallmark of critical phenomena — the slowing down of all processes — leads also to the slowing-down of the rates of chemical reactions. This phenomenon is described in Sec. 2.3, while the opposite peculiar phenomenon of speeding-up of a chemical reaction near the critical point provides the subject matter for Sec. 2.4. It is shown that all three types of behavior (slowing-down, speeding-up and unchanged) are possible depending on the given experiment. In Sec. 2.5, we consider another influence of criticality — the anomalies in chemical equilibria including supercritical extraction. The appropriate experiments are described in Sec. 2.6.

The reverse process — influence of chemistry on critical phenomena — forms the content of Chapter 3, including the change in the critical parameters (Sec. 3.1), critical indices (Sec. 3.2), transport coefficients (Sec. 3.5) and degree of dissociation (Sec. 3.3). Section 3.4 is devoted to the isotope exchange reaction in near-critical systems.

Chapter 4 deals with the problem of the phase separation in reactive systems. The occurrence of multiple solutions of the law of mass action is described in Sec. 4.1. The mechanism of phase separation depends on whether it starts from a metastable or a non-stable state. The former case, where phase separation takes place through nucleation, and the spinodal decomposition for the latter case are considered in Secs. 4.2 and 4.3, respectively. Section 4.4 is devoted to the special case of a dissociation reaction in a ternary mixture.

Chapter 5 contains a description of chemical reactions near some specific regions of the phase diagram. An account of the supercritical fluids is given in Sec. 5.1, while the vicinities of the azeotrope, melting and double critical points are considered in Secs. 5.2, 5.3 and 5.4, respectively. The main experimental methods of analysis of near-critical fluids — sound propagation and light scattering — are considered in Chapter 6. Finally, in Chapter 7 we present our conclusions.

Contents

Chapter 1

Criticality and Chemistry

1.1 Critical phenomena

Phase transitions occur in Nature in a great variety of systems and under a very wide range of conditions. For instance, the paramagnetic-ferromagnetic transition occurs in iron at around 1000 K, the superfluid transition occurs in liquid helium at 2.2 K, and Bose-Einstein condensation occurs at 10^{-7} K. In addition to this enormous temperature range, phase transitions occur in a wide variety of substances, including solids, classical fluids and quantum fluids. Therefore, phase transitions are a very general phenomenon, associated with the basic properties of many-body systems. The thermodynamic functions become singular at the phase transition points, and these mathematical singularities lead to many unusual properties of the system which are called "critical phenomena." We first consider the different types of the phase transition points ("critical points") and then we introduce a qualitative method for describing the behavior of various parameters of the system in the vicinity of critical points.

The liquid-gas critical point of an one-component fluid is determined by the condition [4]

$$\left(\frac{\partial p}{\partial \rho}\right)_T = \left(\frac{\partial^2 p}{\partial \rho^2}\right)_T = 0 \tag{1.1}$$

where p is the pressure, ρ is the density, and T is the temperature. Similarly, the liquid-gas critical points of binary mixtures are characterized by the vanishing of the first and second derivatives of the chemical potential μ with respect to the concentration x,

$$\left(\frac{\partial \mu}{\partial x}\right)_{T,p} = \left(\frac{\partial^2 \mu}{\partial x^2}\right)_{T,p} = 0. \tag{1.2}$$

Here $\mu = \mu_1/m_1 - \mu_2/m_2$, where μ_1, μ_2 and m_1, m_2 are the chemical potentials and masses of the two components.

The close relation between (1.1) and (1.2) is evident from the equivalent form of Eq. (1.2), which can be rewritten as

$$\left(\frac{\partial p}{\partial \rho}\right)_{T,\mu} = \left(\frac{\partial^2 p}{\partial \rho^2}\right)_{T,\mu} = 0. \tag{1.3}$$

The critical conditions for a binary mixture (1.3) are the same as those for a pure system (1.1) when the chemical potential is kept constant. Analogously, the critical points for an n-component mixture are determined by the conditions

$$\left(\frac{\partial p}{\partial \rho}\right)_{T,\mu_1,\ldots,\mu_{n-1}} = \left(\frac{\partial^2 p}{\partial \rho^2}\right)_{T,\mu_1,\ldots,\mu_{n-1}} = 0 \tag{1.4}$$

where $n-1$ chemical potentials are held constant.

In addition to the above-mentioned thermodynamic peculiarities, relaxation processes slow down near the critical points resulting in singularities in the kinetic coefficients. An example is the slowing-down of diffusion near the critical points of a binary mixture. Nothing happens to the motion of the separate molecules when one approaches the critical point. It is the rate of equalization of the concentration gradients by diffusion which is reduced near the critical points. In fact, the excess concentration δx in some part of a system does not produce diffusion by itself. Usually a system has no difficulty in "translating" the change in concentration into a change in the chemical potential $\delta\mu \sim (\partial\mu/\partial x)\,\delta x$, which is the driving force for diffusion. However, near the critical point, according to (1.2), $\partial\mu/\partial x$ is very small, and the system becomes indifferent to changes in concentration. This is the simple physical explanation of the slowing-down of diffusion near the critical point.

Since the states of a one-component system and a binary mixture are defined by the equations of state $p = p\,(T,\mu)$ and $\mu = \mu\,(T,\rho,x)$, respectively, Eqs. (1.1) and (1.2) define the isolated critical point for an one-component system, and the line of critical points for a binary mixture. Another distinction between one-component systems and binary mixtures is that there are two types of critical points in the latter: the above considered liquid-gas critical points and liquid-liquid critical points, whereas two coexisting liquid phases are distinguished by different concentrations of the components. Both critical lines are defined by Eq. (1.2).

Different binary liquid mixtures show either concave-down or concave-up coexistence curves in a temperature-concentration phase diagram (at fixed pressure) or in a pressure-concentration phase diagram (at fixed temperature). The consolute point is an extremum in the phase diagram where the homogeneous liquid mixture first begins to separate into two immiscible liquid layers. For the concave-down diagram, as for a methanol-heplan mixture, the minimum temperature above which the two liquids are miscible in all proportions is called the upper critical solution temperature (UCST). By contrast, for a concave-up diagram, such as the water-triethylamine solution, the maximum temperature below which the liquids are miscible in all proportions is called the lower critical solution temperature (LCST). Under the assumption of analyticity of the thermodynamic functions at the critical points, one can obtain the general thermodynamic criterion for the existence of UCST and LCST [3]. We will discuss this calculation in the next section when examinating the influence of chemical reactions on UCST and LCST. Here we will consider chemical reactions occurring between solutes near the critical points of the solvent.

The properties of near-critical fluids range between those of gases and liquids (see Table 1). Near-critical fluids combine properties of gases and liquids. Their densities are lower than those of liquids, but much higher than the densities of gases, which makes the near-critical fluids excellent solvents for a variety of substances.

Table 1. Comparison of some physical properties of gases, liquids and near-critical fluids.

Physical Properties	Gas	Near-critical fluid	Liquid
Density (kg/m^3)	0.6–2	200–500	600–1000
Kinematic viscosity ($10^{-6}m^2/sec$)	5–500	0.02–0.1	0.1–5
Diffusion coefficient (10^{-6} m^2/sec)	10–40	0.07	2×10^{-4}–2×10^{-3}

In Table 2 we list the critical parameters of the solvents in most common use. Water is the most abundant, cheap, safe and environmentally pure solvent. In spite of its high critical parameters which limits its application, in addition to the traditional uses, modern applications include the important problems of solving the environmental pollution problem and the fabrication of nanocrystalline materials with predictable properties [5]. Properties of near-critical water, such as the full mixing with oxygen and organic compounds, high diffusion and mass transfer coefficients, make water appropriate for efficient treatment of industrial wastes. The use of near-critical water for detoxification of organic waste using the catalytic oxidation of pyridine was found [6] to be cheaper than other methods and also more effective, having almost no limitation on the concentration of the pyridine-containing solutions. The efficiency of hydrothermal detoxification of pyridine waste is substantially increased by the addition of a small amount of heterogeneous catalyst. For instance, the addition of 0.5% of $PtAl_2O_3$ increases the oxidation of pyridine to 99% [7]. Other methods include dechlorination of chlorinate organic compounds, cleaning of polymers and plastic wastes, hydrolysis of cellulose, and the release of bromine for polymers and plastics.

Table 2. Critical parameters of fluids which are commonly used as solvents for chemical reactions.

Solvent	T_{cr} (C)	p_{cr} (atm)	ρ_{cr} (g/mL)
Water (H_2O)	373.9	220.6	0.322
Carbon dioxide (CO_2)	30.9	72.9	0.47
Sulfur hexafluoride (SF_6)	45.5	36.7	0.73
Ammonia (NH_3)	132.3	113.5	0.235
Methanol (CH_3OH)	239.4	80.9	0.272
Propane (C_3H_6)	96.6	41.9	0.22
Ethane (C_2H_6)	32.2	48.2	0.20
Pyridine (C_5H_5N)	347	55.6	0.31
Benzene (C_6H_6)	289	48.3	0.30

Nanocrystallines (particles whose size is a few interatomic distances) are new generation materials widely used as sensors, fuel cells, high-density ceramics, and semiconductors, among others. Hydrothermal synthesis in near-critical water is used to obtain nanocrystalline oxide powders with specified particle sizes and phase composition. Many references can be found [5] dealing with both nanotechnology and environmental problems.

Like water, carbon dioxide (CO_2) has the advantage of being nonflammable, nontoxic and environment compatible. At the same time, CO_2 has critical parameters more convenient than water, and is, therefore, the first choice for use as a near-critical solvent. Another advantage of CO_2 lies in the fact that it does not attack enzymes and is therefore suitable for enzyme-catalyzed reactions. Some new applications include the use of two-phase reaction mixtures with high pressure carbon dioxide which are known as "CO_2 — expanded fluids". In fact, near-critical carbon dioxide is more frequently used in the laboratory and in technology than any other solvent. Hundreds of examples can be found in recent reviews [8], [9].

The infinite increase of the compressibility $\rho^{-1}(\partial\rho/\partial p)_T$ or $(\partial x/\partial\mu)_{T,p}$ as the critical point is approached, leads to a number of peculiarities in the behavior of a substance near its critical point. The specific heat at constant pressure C_p and the expansion coefficient $\beta = -\rho^{-1}(\partial\rho/\partial T)_p$ also increase near the critical point of a one-component system, as follows from Eq. (1.1) and the appropriate thermodynamic relations.

A sharp increase in the mean square fluctuations of the density (or concentration) and of the integral of the correlation function $g_{\rho\rho}$ follows from the well-known thermodynamic relations

$$\overline{(\rho(r)-\overline{\rho})^2} \sim \left(\frac{\partial\rho}{\partial p}\right)_T \to \infty; \quad \int g_{\rho\rho} d^3r \sim \left(\frac{\partial\rho}{\partial p}\right)_T \to \infty. \tag{1.5}$$

The large increase of the correlations between the positions of different particles is given by the second expression in (1.5), which is closely connected with the first expression. In other words, widely separated particles have to be strongly correlated to cause great changes in density.

The correlation radius ξ, which characterizes the distance over which correlations are significant, increases sharply near the critical temperature T_C,

$$\xi_{T\to T_C} \to \infty. \tag{1.6}$$

According to estimates from scattering experiments, ξ reaches $10^{-4} - 10^{-5}$ cm near the critical point. Thus, the specific nature of the critical region consists of the appearance of a new characteristic distance ξ, satisfying the condition

$$a << \xi << R \tag{1.7}$$

where a is the average distance between particles and R is a characteristic macroscopic length.

As an illustration of the crucial importance of the new characteristic length ξ, let us consider the singular part of the transport coefficients near the critical point for a model fluid consisting of spheres with a characteristic radius ξ. Particles inside such spheres are strongly correlated, and we can assume that under the influence of an external force, they move together with a mean velocity v and a mean free path ξ. One finds the following results [10]:

1. **Diffusion coefficient** [11]. When an external force F is applied, the spheres move according to Stoke's law, $F \sim \eta\xi v$, where η is the viscosity, i.e., the mobility $b \equiv v/F \sim (\eta\xi)^{-1}$. Using the Einstein relation $D = k_B T b$,

where D is the diffusion coefficient and k_B is the Boltzmann constant, we have $D \sim (\eta\xi)^{-1}$ or $D\eta \sim \xi^{-1}$, a result confirmed by the more rigorous theory and by experiment.

2. **Heat conductivity**. The usual arguments of molecular-kinetic theory give the heat flux q passing through unit area per unit time, $q \sim vn(\epsilon_1 - \epsilon_2)$. Here, n is the total number of spheres, and $\epsilon_1 - \epsilon_2$ is the difference in their energies on two sides of a selected area, arising from the temperature difference $T_1 - T_2 \sim \xi\nabla T : \epsilon_1 - \epsilon_2 \sim VC_p\xi\nabla T$, where V is the total volume of the spheres, so that $nV = 1$. Thus, $q \sim vC_p\xi\nabla T$. One can find the velocity v from the estimate for the diffusion coefficient given above: $v \sim D/\xi \sim 1/\xi^2\eta$. Finally, the heat conductivity $\lambda \sim q/\nabla T \sim vC_p\xi \sim C_p/\eta\xi$. This result is supported by more rigorous theory and also by experiment.

The qualitative description of the singularity of the quantity a at the critical point is given by the non-integer critical index x, where $a \sim |T - T_C|^x$. If $x \neq 0$, this critical index is given by $x = \ln(a)/\ln|T - T_C|$, whereas if $x = 0$, there are two possibilities. In one case, a becomes constant at the critical point, with the possibility of different values of this constant on the two sides of the transition, which is called a jump singularity. In the other case, a exhibits a logarithmic singularity, $a \sim \ln|T - T_C|$.

Although there are general properties which define the values of the critical indices (dimension d of space, symmetry and the presence of long-range interactions), according to the universality principle, these values are defined by the general statistical properties of the many-body system, rather than by the details of the microscopic interactions. This implies that different isomorphic systems will have the same indices for the appropriate parameters. Using the language of fluids, the commonly accepted symbols $\alpha, \beta, \gamma, \delta, \eta$, and ν describe, respectively, the asymptotic behavior near the one-component liquid-gas critical point of the specific heat at constant volume, order parameter (deviation of the density from its critical value), compressibility, pressure-density relation at the critical isotherm, the correlation function at the critical temperature, and the correlation length. Of these indices, one (β) is determined only below the critical temperature and two (δ, η) exist only at the critical temperature. The remaining three indices can be defined for temperatures above critical (α, γ, ν) as well as below (α', γ', ν'). According to scaling theory [12], $\alpha = \alpha', \gamma = \gamma'$, $\nu = \nu'$ and the following relations exist between critical indices: $\alpha + 2\beta + \gamma = 2$, $d\nu = 2 - \alpha$, $(2 - \eta)\nu = \gamma$, and $\beta(\delta - 1) = \gamma$.

The values of the critical indices depend on the proximity to the critical point. In the approach to the critical point, the indices have their "classical" mean-field values, obtained by assuming the analyticity of thermodynamic functions at the critical points which allows the expansion of these functions in a power series in the deviation from the critical parameters. However, the singularity in the critical points makes the series expansion inapplicable very close to the critical point, leading to "non-classical" values of the critical indices (non-integer powers, logarithmic dependence, etc.). Since the difference between the results of mean-field theory and the exact theory is due to fluctuations of the order parameter, we expect that the mean-field approximation will be accurate when these fluctuations are small. Ginzburg proposed [13] that the mean field theory is applicable when the fluctuations are small compared to the thermodynamic values. The Ginzburg criterion divides the region near the critical point into two parts, giving the crossover from classical to non-classical behavior. An interesting idea has been proposed recently [14] of the existence of the second crossover in the immediate vicinity of the critical point, where the non-classical critical indices regain their classical values. Indeed, according to Eq. (1.5), the development of large-scale fluctuations is accompanied by a continuous increase of the sus-

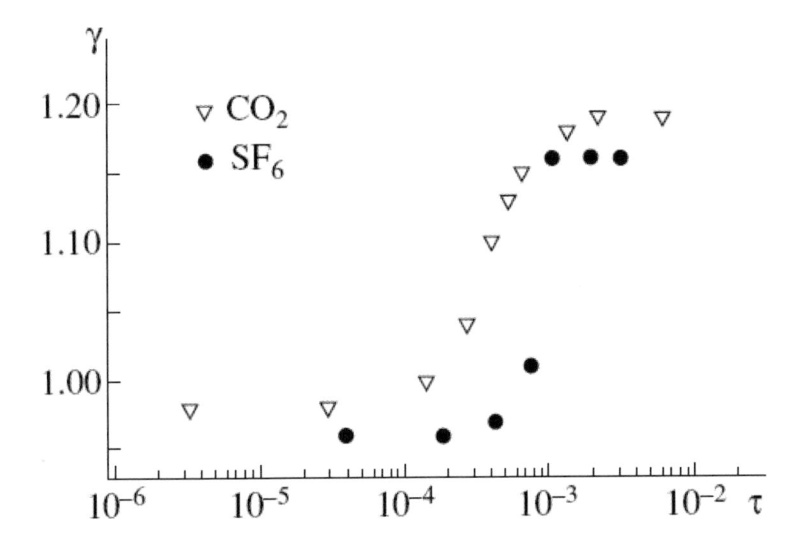

Fig. 1.1 Variation of critical exponent for the isothermal compressibility in the immediate vicinity of the critical point for CO_2 and SF_6. The experimental points are taken from [16], [17]. Reproduced from Ref. [14] with permission, copyright (2002), Springer.

ceptibilities of the critical system, in particular, the susceptibility to varies external perturbations (gravitational and Coulomb fields, surface forces, shear stresses, turbulence, presence of boundaries). As a consequence, classical mean-field behavior must be restored. This transition occurs in the direction opposite to the Ginzburg criterion direction and defines the second crossover. Such behavior was found experimentally as early as 1974 [15] by taking $p - V - T$ measurements of very pure SF_6 in the immediate vicinity ($\tau = \frac{T-T_C}{T_C}$) of the critical point with 2×10^{-4}K, $\pm 0.01\%$, and $\pm 0.02\%$ accuracy, respectively in the temperature, dimensionless pressure and density.

As an example, we show [14] in Fig. 1.1 the crossover to the classical value of the critical index of the isothermal compressibility in the immediate vicinity of the critical point for CO_2 and SF_6. The author of [14] states his belief that the analogous second crossover has been seen in other experiments under the influence of gravity, impurities, and shear flows. This problem certainly deserves further investigation.

1.2 Chemical reactions

The equilibrium numbers of particles of the different substances taking part in a chemical reaction are connected by the law of mass action. This law results from a relation between the chemical potentials μ_i of the various components [3]. Thus, for the reaction $\sum \nu_i A_i = 0$, where A_i are the chemical symbols of the reagents and ν_i are positive or negative integers, the equation for chemical equilibrium has the form $\sum \nu_i \mu_i = 0$. For the simplest case of the isomerization reaction, the reaction equation and the law of mass action have the form $A_1 - A_2 = 0$ and $\mu_1 = \mu_2$, respectively.

Thus, binary mixtures undergoing a chemical reaction are characterized by their concentration, as in a non-reactive mixture. However, according to the law of mass action, this concentration is a function of other thermodynamic variables. Therefore, the number of thermodynamic degrees of freedom of a binary system undergoing a chemical reaction is the same as for a non-reactive, one-component system.

Chemical reactions influence all properties of many-component systems. As an example, we shall show that the existence of the chemical reaction may lead to the replacement of UCST by LCST and vice versa [18]. Moreover, the Clapeyron-Clausius equation for a binary mixture is determined by the chemical reaction, in addition to the latent heat and the volume difference between the two phases.

Consider a mole of a binary mixture which separates into two phases, B' and B''. The system can be described by four parameters, p, T, x_2', and x_2'', which satisfy the following conditions:

$$\mu_1'(T,p,x_2') = \mu_1''(T,p,x_2''); \quad \mu_2'(T,p,x_2') = \mu_2''(T,p,x_2''). \tag{1.8}$$

One can differentiate the equilibrium conditions (1.8) along the equilibrium surface between B' and B''. Using simple thermodynamic relations, one obtains [3]

$$\Delta v_1 dp - \Delta h_1 dT/T + x_2' g_{2x}' dx_2' - x_2'' g_{2x}'' dx_2'' = 0,$$

$$\Delta v_2 dp - \Delta h_2 dT/T + (1 - x_2'') g_{2x}'' dx_2'' - (1 - x_2') g_{2x}' dx_2' = 0 \tag{1.9}$$

where v_i and h_i are the partial molar volume and enthalpy,

$$\Delta v_i = v_i'' - v_i'; \; \Delta h_i = h_i'' - h_i'; \; g_{2x}' \equiv \left(\frac{\partial^2 g'}{\partial x_2'^2}\right)_{T,p}; \; g_{2x}'' \equiv \left(\frac{\partial^2 g''}{\partial x_2''^2}\right)_{T,p}. \tag{1.10}$$

The partial derivatives of the intensive variables are given by Eq. (1.9) at constant pressure (temperature) and at constant concentration [3]. However, there is no need to consider these special sections of the coexistence surface when we deal with a reactive system. Indeed, for a reaction $\nu_1 A_1 \rightleftarrows \nu_2 A_2$, an additional restriction to (1.9) exists in the form of the law of mass action

$$\nu_1 \mu_1' + \nu_2 \mu_2' = 0. \tag{1.11}$$

Differentiating the latter equation along the equilibrium surface, yields

$$\begin{aligned}(\nu_1 v_1' + \nu_2 v_2')\, dp - (\nu_1 h_1' + \nu_2 h_2')\, dT/T \\ + [\nu_1 x_2' g_{2x}' - \nu_2 (1 - x_2') g_{2x}']\, dx_2' = 0.\end{aligned} \tag{1.12}$$

Combining Eqs. (1.9) and (1.12) yields the slope of the equilibrium line of a two-phase reactive binary mixture,

$$T\left(\frac{\partial p}{\partial T}\right)_{chem} = \frac{h_{2x}' (\Delta x_2)^2 - (2\Delta x_2/n')(\nu_1 v_1' + \nu_2 v_2')}{v_{2x}' (\Delta x_2)^2 - (2\Delta x_2/n')(\nu_1 h_1' + \nu_2 h_2')} \tag{1.13}$$

$$\left(\frac{\partial T}{\partial x_2'}\right)_{chem} = \frac{2T g_{2x}' \Delta x_2 + T n' v_2' (\Delta x_2)^2 (\nu_1 v_1' + \nu_2 v_2')^{-1} g_{2x}'}{h_{2x}' (\Delta x_2)^2 - v_2' (\Delta x_2)(\nu_1 h_1' + \nu_2 h_2')(\nu_1 v_1' + \nu_2 v_2')^{-1}} \tag{1.14}$$

where $n' \equiv \nu_1 x_2' - \nu_2 (1 - x_2')$ and $h' \equiv (1 - x_2') h_1' + x_2' h_2'$; $v' \equiv (1 - x_2') v_1' + x_2' v_2'$ are the heat of reaction and the volume change of reaction in phase B'.

In the absence of a chemical reaction, all terms vanish, except the first term in the denominators and numerators of Eqs. (1.13) and (1.14), and these terms reduce to the well-known form [3]

$$T\left(\frac{\partial p}{\partial T}\right)_{x_2'} = \frac{h_{2x}'}{v_{2x}'} \tag{1.15}$$

$$\left(\frac{\partial T}{\partial x_2'}\right)_p = \frac{2Tg_{2x}'\Delta x_2}{h_{2x}'\left(\Delta x_2\right)^2}. \tag{1.16}$$

Equation (1.13) is the generalized form of the Clapeyron-Clausius equation (1.15) for a reactive mixture, while Eq. (1.14) determines the criterion for UCST and LCST. The latter can be obtained in the same way as for a nonreactive mixture [3].

The "classical" expansion near the critical points $g_{2x}' \approx \frac{1}{8}g_{4x}'\left(\Delta x_2\right)^2$, yields

$$\left(\frac{\partial T}{\partial x_2'}\right)_p = \frac{Tg_{4x}'\left(x_2' - x_2''\right)}{4h_{2x,cr}}. \tag{1.17}$$

If $x_2'' > x_{2,cr} > x_2'$, then $\left(\partial x_2'/\partial T\right)_p$ is positive if $h_{2x,cr} \equiv \left(\partial^2 h/\partial x^2\right)_{cr}$ is negative. These signs define UCST. Analogously, LCST corresponds to $\left(\partial^2 h/\partial x^2\right)_{cr} > 0$.

Performing a similar expansion near the single critical point of a reactive binary mixture, one obtains from (1.13) and (1.14),

$$T\left(\frac{\partial p}{\partial T}\right)_{chem} \sim \frac{v_1 h_{1,cr} + v_2 h_{2,cr}}{v_1 v_{1,cr} + v_2 v_{2,cr}} \tag{1.18}$$

$$\begin{aligned}\left(\frac{\partial T}{\partial x_2'}\right)_{chem} &\sim Tg_{4x}'\left(x_2' - x_2''\right)\left(h_{2x,cr} - v_{2x,cr}\frac{v_1 h_{1,cr} + v_2 h_{2,cr}}{v_1 v_{1,cr} + v_2 v_{2,cr}}\right)^{-1} \\ &\sim Tg_{4x}'\left(x_2' - x_2''\right)h_{2x,cr}\left(1 - \frac{v_1 h_{1,cr} + v_2 h_{2,cr}}{T_c\left(v_1 v_{1,cr} + v_2 v_{2,cr}\right)}dT_C/dp\right)^{-1}.\end{aligned} \tag{1.19}$$

Equation (1.15) was used in the last relation in (1.19). It follows from (1.19) that the existence of a chemical reaction may change the type of critical point (UCST to LCST and vice versa) if the last bracket in (1.19) is negative. The ratio of the second derivatives $v_{2x,cr}/h_{2x,cr}$ can be replaced by the ratio of excess volume V_E and excess enthalpy h_E at the critical point.

Thus, a chemical reaction will change the nature of the critical point if the following (equivalent) inequalities are satisfied:

$$\frac{v_1 h_{1,cr} + v_2 h_{2,cr}}{v_1 v_{1,cr} + v_2 v_{2,cr}} \frac{V_E}{h_E} > 1 \text{ or } \frac{v_1 h_{1,cr} + v_2 h_{2,cr}}{v_1 v_{1,cr} + v_2 v_{2,cr}} \frac{1}{T_C} \frac{dT_C}{dp} > 1. \quad (1.20)$$

One can find [19], [20] the form of the critical line near the critical point dT_C/dp as well as V_E and h_E. Typical examples of positive V_E and h_E are shown in Fig. 1.2.

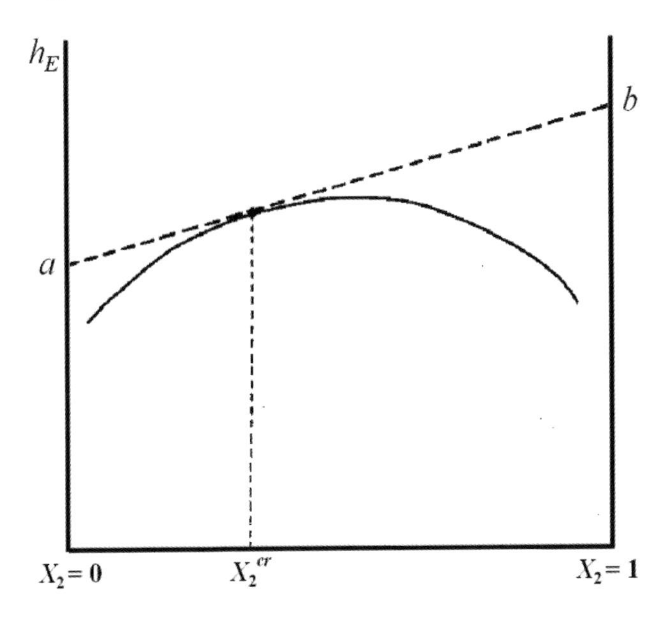

Fig. 1.2 Typical form of the excess enthalpy h_E (or excess volume V_E) for a binary mixture as a function of concentration. Points a and b correspond to the critical molar enthalpy of pure substances. Reproduced from Ref. [18] with permission, copyright (1990), Springer.

Using the definition of the partial molar quantity $y_1 = y - x_2 \left(\partial y/\partial x_2\right)_{T,p}$, where $y \equiv (h, v)$, one can see that the points a, b in Fig. 1.2 give $h_{1,cr}$ and $h_{2,cr}$ (or, analogously, $v_{1,cr}$ and $v_{2,cr}$). For an isomerization reaction, $\nu_1 = -\nu_2 = 1$, and Eq. (1.20) becomes

$$\frac{h_{1,cr} - h_{2,cr}}{h_E} \frac{V_E}{v_{1,cr} - v_{2,cr}} > 1. \quad (1.21)$$

There is no physical reason why criterion (1.21) should not be satisfied for some mixtures. Then, the presence of a chemical reaction will change UCST to LCST and vice versa.

1.3 Analogy between critical phenomena and the instability of chemical reactions

Critical phenomena describe the behavior of closed thermodynamic systems whereas chemical reactions occur in the systems open to matter transport from the environment. The former are described by the well-known Gibbs technique, but there is no universal approach to non-equilibrium chemical reactions which are defined by the equation of the reaction rate. However, as already mentioned in the Introduction, there is a close analogy between these two phenomena (in fact, article [21] is entitled "Chemical instabilities as critical phenomena").

One distinguishes between phase transitions of first and second orders. First-order transitions involve a discontinuity in the state of the system, and, as a result, a discontinuity in the thermodynamic variables such as entropy, volume, internal energy (first derivatives of the thermodynamic potential). In second-order phase transitions, these variables change continuously while their derivatives, which are the second derivatives of the thermodynamic potentials (specific heat, thermal expansion, compressibility), are discontinuous. Analogously, in the theory of instability of nonlinear differential equations, which describe the rate of chemical reactions, one distinguishes between hard and soft transitions. These are similar in nature to first-order and second-order phase transitions [22].

As an example, consider the following chemical reactions [23]

$$A + 2X \underset{k_2}{\overset{k_1}{\rightleftarrows}} 3X; \qquad A \underset{k_4}{\overset{k_3}{\rightleftarrows}} X. \tag{1.22}$$

These rate equations describe the conversion of the initial reactant A into X by two parallel processes: a simple monomolecular degradation or an autocatalytic trimolecular reaction. Both these reactions are reversible with reaction constants k_i, $i = 1, \ldots, 4$. The system is open to interaction with an external reservoir of reactant A, so that the concentration of A remains constant. The macroscopic equation for the number of molecules X has the following form

$$\frac{dX}{dt} = -k_2 X^3 + k_1 A X^2 - k_4 X + k_3 A. \tag{1.23}$$

The solution of Eq. (1.23) with the initial condition $X(0) = X_0$ is

$$\left(\frac{X - X_1}{X_0 + X_1}\right)^{k_3 - k_2} \left(\frac{X - X_2}{X_0 - X_2}\right)^{k_1 - k_3} \left(\frac{X - X_3}{X_0 - X_3}\right)^{k_2 - k_1}$$
$$= \exp\left[-k_2 (X_1 - X_2)(X_2 - X_3)(X_3 - X_1)\, t\right] \tag{1.24}$$

where X_1, X_2 and X_3 are the three roots of

$$k_2X^3 - k_1AX^2 + k_4X - k_3A = 0 \tag{1.25}$$

with $X_3 \geq X_2 \geq X_1$. The steady state solutions X_s of Eq. (1.24) are

$$\begin{aligned} X_s &= X_1 \text{ for } X_0 < X_2; \\ X_s &= X_2 \text{ for } X_0 = X_2; \\ X_s &= X_3 \text{ for } X_0 > X_2. \end{aligned} \tag{1.26}$$

Stability analysis shows that the solution X_2 is unstable with respect to small perturbations, whereas the solutions X_1 and X_3 are stable. Moreover, it follows from Eq. (1.26) that hysteresis may occur as X is varied [23]. The last two results are typical of equilibrium phenomena which are described near the liquid-gas critical point by the classical equation of state, say, the van der Waals equation of the form (1.25). This establishes the link between first-order phase transitions in equilibrium systems and so-called hard transitions in reactive systems.

As an example of different behavior, consider the chemical reaction

$$A + C + X \rightleftarrows A + 2X; \qquad X \to Y + B \tag{1.27}$$

where the second reverse reaction is neglected. The rate equation for the number of molecules X has the following form

$$\frac{dX}{dt} = -AX^2 + (AC - 1)\,X \tag{1.28}$$

or, introducing $\tau = At$ and $\lambda = (AC - 1)\,/A$,

$$\frac{dX}{d\tau} = -X^2 + \lambda X. \tag{1.29}$$

The solution of this equation has the typical features of second-order phase transitions: the soft transition points $X_s\,(\lambda)$ are continuous at the transition point $\lambda = 0$, but the derivatives are not.

The foregoing equations have to be generalized to include fluctuations from the steady state. No first-principle microscopic theory exists for fluctuations in reactive chemical systems. One usually uses a phenomenological master equation based on the macroscopic rate equations or a Langevin equation obtained by adding a stochastic term to the rate equations. The details can be found in [22]. Here we bring a fascinating example of the influence of noise on the chemical reaction [24], which is illustrated by the so-called ecological model. In this model, one considers two biological species with densities $n_1\,(t)$ and $n_2\,(t)$, which decrease with rates a_1 and

a_2 and compete with rates b_1 and b_2 for the same renewable food m. The birth-death equations for these species have the following form

$$\frac{dn_1}{dt} = (b_1 m - a_1)\, n_1; \qquad \frac{dn_2}{dt} = (b_2 m - a_2)\, n_2. \tag{1.30}$$

The amount of food decreases both naturally (at a rate c), and according to Eq. (1.30). Assume that the amount m of the food increases at rate q, i.e.,

$$\frac{dm}{dt} = q - cm - d_1 n_1 - d_2 n_2. \tag{1.31}$$

Assume now that the species n_1 is strong and species n_2 is weak, which occurs when the ratio a_1/b_1 is smaller than both a_2/b_2 and q/c. Under these conditions, the asymptotic $t \to \infty$ solutions of Eqs. (1.30) and (1.31) are

$$m = \frac{a_1}{b_1}; \qquad n_2 = 0; \qquad n_1 = \frac{qb_1 - ca_1}{b_1 d_1} \tag{1.32}$$

which means that in the long run, the strong species survives and the weak species becomes extinct. The question arises regarding which changes can help the weak species to survive. One can easily see [24] that if one allows the food growth rate q to fluctuate in time (replacing q in Eq. (1.31) by $q + f(t)$ with $\langle f \rangle = 0$) or to allow the weak (but not the strong!) species to be mobile (adding the diffusive term $D\nabla^2 n_2$ term in the second of Eqs. (1.30)), the weak species will finally become extinct. This result can be formulated as the "ecological theorem": the long-term coexistence of two species relying on the same renewable resources is impossible. However, it has been shown [24] that if both these factors occur together with not-too-small food growth fluctuations, the corrected equations (1.30) and (1.31) have non-zero asymptotic solutions for both n_1 and n_2. This result has an important ecological interpretation. In a fluctuating environment, high mobility gives an evolutionary advantage that makes possible the coexistence of weak and strong species. The individuals of the mobile weak species survive since they can utilize the food growth rate fluctuations more effectively.

The effect outlined above, which is called a "noise-induced phase transition" [25], also occurs in systems undergoing chemical reactions [24], as well as in many other phenomena.

Chapter 2

Effect of Criticality on Chemical Reaction

2.1 The effect of pressure on the equilibrium constant and rate of reaction

Dozens of experimental results for different chemical reactions (solutes) occurring in different near-critical substances (solvents) have been reported in the chemical literature. One such example is the tautomeric equilibrium in supercritical 1,1-difluoroethane [26], where the equilibrium rate constant increased four-fold for a small pressure change of 4 MPa. Such a substantial increase of the reaction rate by small variations of pressure has also been found [27] in styrene hydroformylation in supercritical CO_2. One further example is the decomposition reaction of 2,2-isobutyronitrile in mixed CO_2-ethanol solvent [28], where a 70–80% increase of the reaction rate near the critical point resulted from a small pressure change, whereas outside the critical region, the reaction rate is nearly independent of pressure.

There are many experimental data concerning the Diels-Alder reaction [29], [30], one of the most interesting and useful reactions in organic chemistry and often used for the synthesis of six-membered rings. The discoverers of this reaction were awarded the Nobel Prize in Chemistry in 1950 "for their discovery and development of diene synthesis". In Wikipedia this reaction is considered the "Mona Lisa" of reactions in organic chemistry since it requires very little energy to create very useful cyclohexane rings. The Deals-Alder reaction has been studied both near the liquid-gas critical point of water, CO_2 and SF_6, as well as near the critical point of binary mixtures, such as CO_2 + ethanol or CO_2 + hexane. In Fig. 2.1, we show [30] the dramatic increase of the reaction rate of the CO_2+ ethanol mixture as the pressure approaches the critical point for composition.

From a technological point of view, the CO_2 + ethanol mixture has the

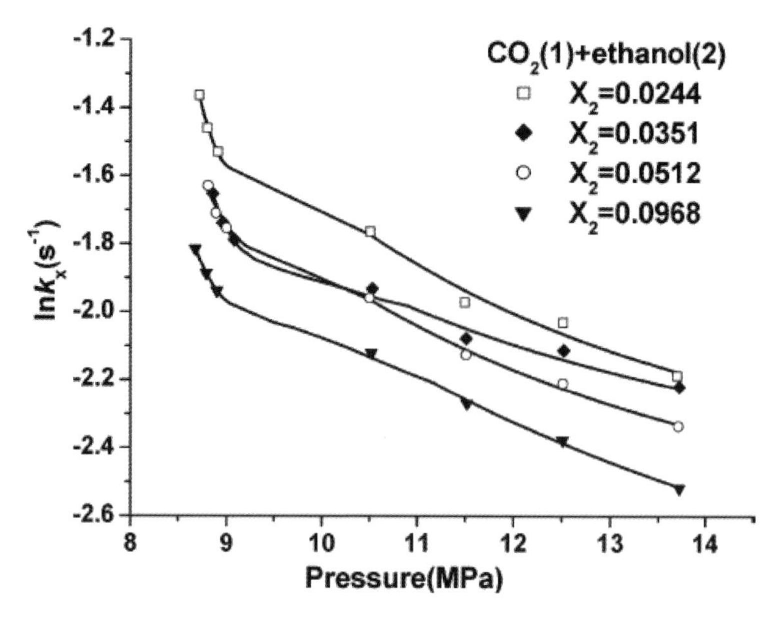

Fig. 2.1 Rate of reaction as a function of pressure for CO_2-ethanol mixture at T = 318.15 K near the critical points. Reproduced from Ref. [30] with permission, copyright (2005), American Chemical Society.

advantage that the critical points are located at low pressures and temperatures. The measurement of the reaction rate of an isoprene-maleic anhydride reaction near both the liquid-gas and liquid-liquid critical points of the hexane-nitrobenzene mixture demonstrates [31] anomalous enhancement near the liquid-gas critical point, in contrast to a very small change near the liquid-liquid critical point.

We now consider the theoretical explanation for the strong influence on chemical reactions of changes in pressure. It is not surprising that a small change in the pressure near the liquid-gas critical point of a pure solvent causes a large change in density-dependent properties, such as the solubility parameter, refractive index and the dielectric constant. To account for these facts, it is enough to take into account the huge increase of the compressibility $(\partial\rho/\partial p)_T$ near the liquid-gas critical point. However, the detailed influence on chemical reactions requires a more sophisticated analysis.

To interpret the pressure effect on reaction kinetics, one uses transition-state theory [32], according to which the activated complex C is formed through the bimolecular reaction, expressed as

$$A + B \longleftrightarrow C \rightarrow P \tag{2.1}$$

where A and B are the reactants and P is the product. The pressure dependence of the rate constant K, expressed in mole fraction units, is given by

$$RT\frac{\partial \ln K}{\partial p} = -\Delta V^* \equiv \overline{V_C} - \overline{V_A} - \overline{V_B} \tag{2.2}$$

where ΔV^* is the apparent activation volume created from the partial molar volumes $\overline{V_i}$ of component i. Being negative, partial molar volumes dramatically increase upon approaching the critical points, peaking at values of $-15,000$ cm^3/gmol [33] compared with $5 - 10$ cm^3/gmol far from the critical point. The large values of the partial molar volumes near the critical points of maximum compressibility are evident from the thermodynamic relation

$$\overline{V_i} = vk_T n \left(\frac{\partial p}{\partial n_i}\right)_{T,V,n_{j\neq i}} \tag{2.3}$$

where k_T is the isothermal compressibility of pure solvent, v is the molar volume of the solvent, n_i is the number of moles of component i, and n is the total number of moles of the mixture. Moreover, the limiting value $\overline{V_{cr}}$ of the partial molar volumes at the critical point is path-dependent. When the critical point is approached from above, the coexistence curve of the binary mixture SF_6 -CO_2, $\overline{V_{cr}} = -230$ cm^3/gmol and along the isothermal-isochoric path $\overline{V_{cr}} = -40$ cm^3/gmol, whereas $\overline{V_{cr}}$ of pure SF_6 equals 198 cm^3/gmol.

2.2 Effect of phase transformations on chemistry

One form of the interplay between chemical reactions and phase transformations is the change of a slow reaction in a liquid mixture in which the transition from the two-phase to the one-phase region results from the change of composition during the reaction. The reaction rate was determined [34] by measuring the heat production $w(t)$ as a function of time with a heat flow calorimeter. Thus, the chemical reaction itself serves as a probe for obtaining information about critical phenomena. There are a number of requirements for the selection of a suitable chemical reaction:

1. It must start in the two-phase region and end in the one-phase region or vice versa.

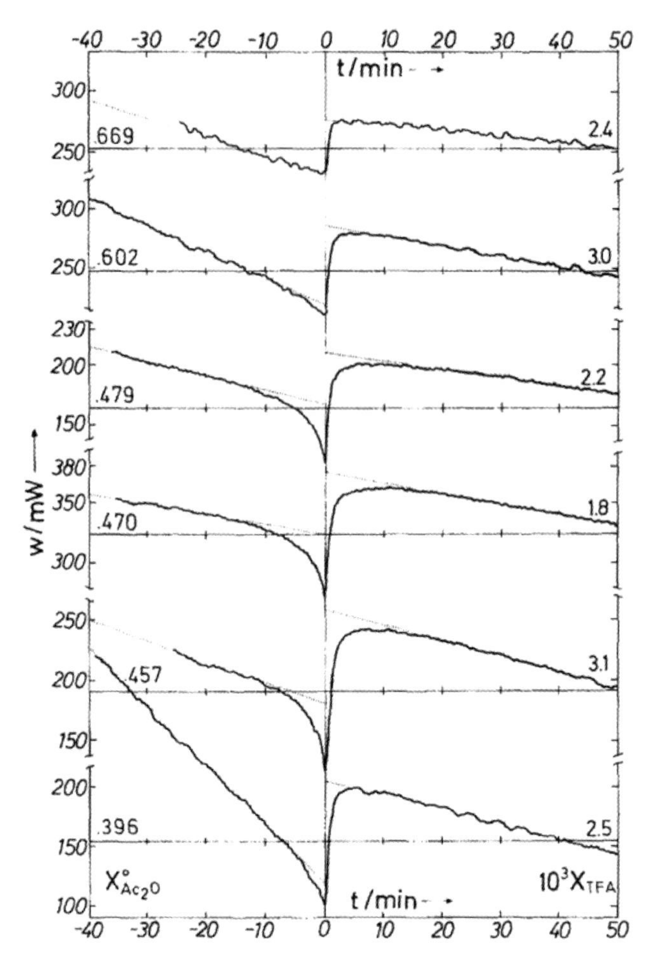

Fig. 2.2 Heat flow curves for the reaction of acetic anhydride with 1, 2-ethanediol with initial mole fractions $x^0_{\mathrm{AC_2O}}$ and x^0_{TFA}, respectively. Reproduced from Ref. [34] with permission, copyright (1989), Elsevier.

2. It must have a slow reaction rate to yield reliable data for the composition of reactants.

3. The reaction enthalpy must be large enough for heat flow calorimetry.

The authors of [34] have chosen the esterification of 1, 2 -ethanediol by acetic anhydride in the presence of a catalytic amount of trifluoroacetic acid. The results are shown in Fig. 2.2, where the different curves relate to different initial amounts of reactants. The minimum heat production is taken as the zero of the time scale so that time t is positive for one-phase regions and negative for two-phase regions.

The curves are characterized by the following features:

1. Far from the phase transition, the curves have negative slope due to the decrease of the reactant concentrations with time.

2. The curves exhibit a 10 – 20% jump in heat production in crossing the frontier between two-phase and one-phase regions.

3. The jumps are almost symmetric with respect to the phase transition point, and fade away as the phase transition occurs more distant from the consolute point.

2.3 Critical slowing-down of chemical reactions

Experimental studies of chemical reactions near the critical point have a long history. Although the general idea of the slowing-down of chemical reactions is well-known [35], theoretical interest in this problem was rekindled in 1981 by our work [36]. For the reaction

$$\sum_{i=1}^{k} \nu_i A_i = \sum_{j=1}^{l} \nu'_j A'_j, \tag{2.4}$$

the observed rate of reaction r is the difference between the forward and backward processes,

$$r = \overrightarrow{r} - \overleftarrow{r} = k_f C_1^{\nu_1} \cdots C_k^{\nu_k} - k_b C_1^{\nu'_1} \cdots C_l^{\nu'_l} \tag{2.5}$$

where k_f and k_b are the microscopically determined rate coefficients for the forward and backward reactions, and C_i are the concentrations (fugacities) of the i-th species in an ideal (nonideal) solution. Using the definition of the chemical potentials of the mole of substance i, $\mu_i = \mu_i^0 + RT \ln C_i$, and the fact that [3]

$$\frac{k_f}{k_b} = \exp \left[\left(\sum_{i=1}^{k} \nu_i \mu_i^0 - \sum_{j=1}^{l} \nu'_j \mu_j^0 \right) / RT \right] \tag{2.6}$$

one can rewrite Eq. (2.5) as

$$\begin{aligned} r &= k_f C_1^{\nu_1} \cdots C_k^{\nu_k} \left[1 - \left(k_b C_1^{\nu_{1'}} \cdots C_l^{\nu_{l'}} \right) \left(k_f C_1^{\nu_1} \cdots C_k^{\nu_k} \right)^{-1} \right] \\ &= k_f C_1^{\nu_1} \cdots C_k^{\nu_k} \left[1 - \exp\left(-A/RT \right) \right], \end{aligned} \tag{2.7}$$

where A is the affinity of reaction.

For chlorine dissociation, $Cl_2 \rightleftarrows 2Cl$, the rate r is defined as $r = k_f C_{Cl_2} - k_b C_{Cl}^2$, and, using $\mu_i = \mu_i^0 + RT \ln C_i$, and $k_f/k_b = \exp \sum \nu_i \mu_i^0 / RT$, one gets

$$r = k_f C_{Cl_2} \left[1 - \exp\left(-\frac{A}{RT}\right)\right]. \tag{2.8}$$

Equation (2.8) implies that, in general, the rate of reaction r is a nonlinear function of A. However, at equilibrium, $A = 0$, and for small deviations from equilibrium, $A/RT << 1$, Eq. (2.7) takes the following form,

$$r = LA = L\left(\frac{\partial A}{\partial \xi}\right)_{T,\ p} (\xi - \xi_{eq}) + \cdots \tag{2.9}$$

where $L = k_f C_1^{\nu_1} \cdots C_k^{\nu_k}/RT$. Therefore, for small deviations from equilibrium, at fixed temperature and pressure, one can restrict ourselves to the first term in the expansion of the affinity $A(T, p, \xi)$ in terms of the extent of reaction ξ.

Another way to obtain Eq. (2.9) is to note that at equilibrium, where the extent of reaction is equal to ξ_{eq}, both $d\xi/dt$ and A vanish, and one gets

$$\frac{d\xi}{dt} = L\left(\frac{\partial A}{\partial \xi}\right)_{T,p} (\xi - \xi_{eq}). \tag{2.10}$$

A detailed analysis of Eq. (2.10) has been performed [37]. Before discussing this work, one needs to recall some general thermodynamic ideas. The thermodynamic state of a one-component, one-phase system is determined by two variables, say, temperature and pressure. Similarly, the state of an n-component, one-phase system is determined by $n+1$ variables. These could be the temperature T and the chemical potential of each component μ_i $(i = 1, 2, \ldots, n)$.

As a function of these "field" variables, the free energy G has the following form

$$G = G(T, \mu_1 \ldots \mu_n). \tag{2.11}$$

However, for a reactive system, the chemical potentials are not independent. According to the law of mass action in equilibrium,

$$A \equiv \sum_{i=1}^{n} \nu_i \mu_i = 0. \tag{2.12}$$

The unconstrained free energy $\overline{G}$ can be written as

$$\overline{G}(T, \mu_1 \ldots \mu_n) = G(T, \mu_1 \ldots \mu_n) + \xi \sum_{i=1}^{n} \nu_i \mu_i \tag{2.13}$$

where ξ is a Lagrange multiplier. By differentiating Eq. (2.13) with respect to μ_i, one obtains the particle number densities

$$n_i = n_i^0 + \nu_i \xi. \tag{2.14}$$

Equation (2.14) implies that for a given system (with initial n_i^0), the change in n_i is completely determined by the extent ξ of the reaction. For given T and p, the latter quantity defines the trajectories of all possible thermodynamic states of a system in n-dimensional phase space. However, the complete stability analysis, at given T and p, requires taking into account all possible changes of n_i coming both from n_i^0 and ξ in Eq. (2.14). In multicomponent systems, the stability with respect to diffusion is broken before the thermal or mechanical stability. In terms of the Gibbs free energy $G\left(T, p, n_1 ... n_n\right)$, the condition for diffusion stability for given T and p has the form

$$d^2 G = \sum_{i,j} dn_i \mu_{ij} dn_j > 0; \qquad \mu_{ij} \equiv \partial \mu_i / \partial n_j. \tag{2.15}$$

One can easily show [3] that the matrix μ_{ij} is semipositive, i.e., the determinant of the matrix μ_{ij} vanishes, which means that the rank of μ_{ij} is equal to $n-1$, and all the eigenvalues of μ_{ij} but one are positive. Let N_j^0 be the eigenvector corresponding to eigenvalue zero,

$$\sum_j \mu_{ij} N_j^0 = 0. \tag{2.16}$$

Equation (2.16) is the Gibbs-Duhem relation, which has the simple physical meaning of invariance with respect to proportional change of all particle numbers. On substituting (2.14) into (2.15), one obtains

$$d^2 G = \sum_{i,j} \left[dn_i^0 \mu_{ij} dn_j^0 + 2 dn_i^0 \mu_{ij} \nu_j d\xi + \nu_i \mu_{ij} \nu_j \left(d\xi\right)^2 \right] > 0. \tag{2.17}$$

The form $q \equiv \sum_{i,j} x_i \mu_{ij} x_j$ vanishes only when the determinant of the matrix μ_{ij} vanishes (however, the opposite is not true, namely, for zero determinant of the matrix, q may remain positive). Therefore, the existence of a chemical reaction does not change the stability conditions. The equivalence of the stability conditions with respect to diffusion becomes evident when one considers [3] these conditions as resulting from homogeneous and nonhomogeneous perturbations of an initially homogeneous system.

On the critical hypersurface, the expression (2.15) vanishes, i.e., new zero eigenvalues of μ_{ij} appear which correspond to the eigenvector N_j satisfying

$$\sum_j \mu_{ij} N_j = 0. \tag{2.18}$$

Among all points on the critical hypersurface, those worthy of notice occur where the condition (2.15) is violated, and also

$$\sum_{i,j} \nu_i \mu_{ij} \nu_j = 0. \tag{2.19}$$

According to Eq. (2.17), this sum equals $(dA/d\xi)_{T,p}$, which, according to Eq. (2.9), determines the slowing-down of chemical reactions.

Let us now return [37] to the main point — the comparison of Eqs. (2.16), (2.18), and (2.19), which shows that the condition (2.19) for critical slowing-down will be satisfied if and only if

$$\nu_i = \alpha N_i + \beta N_i^0 \tag{2.20}$$

where α and β are arbitrary constants. There are $n - 2$ constraints in (2.20), equal to the rank of the matrix μ_{ij} on the critical hypersurface. Together with two critical conditions, they define n conditions on $n + 1$ independent variables. Therefore, there is one free parameter, which is fixed by the law of mass action. As a result, slowing-down may appear only at an isolated point on the critical hypersurface provided that all the constraints are compatible. Each chemical reaction decreases by unity the dimension of the critical hypersurface. For an n-component system with $n - 1$ chemical reactions (for example, a binary mixture with a single chemical reaction), the only existing critical point will be where slowing-down occurs. For an n-component mixture with fewer than $n - 1$ chemical reactions, the chemical instability point will be an isolated point on the critical hypersurface, which makes experimental verification very difficult. However, in addition to a binary mixture with a single chemical reaction, slowing-down is also very probable [37] for reactions in multicomponent systems involving the separation of slightly-dissolved substances from a solvent.

Another attack on the slowing-down problem is the general Griffiths-Wheeler method [38] of estimating the singularity of thermodynamic derivatives at the critical points. Indeed, according to Eq. (2.9), slowing-down is determined by the thermodynamic derivative $(dA/d\xi)_{T,p}$, and the question arises whether this derivative has the singularity at the critical point. Griffiths and Wheeler [38] divide all thermodynamic variables into two classes: "fields" and "densities". The fields (temperature, pressure, chemical potentials of the components) are the same in all phases coexisting in equilibrium, whereas the densities (volume, entropy, concentrations of chemical components, extents of reactions) have different values in each coexisting phase. The power of the singularity of the thermodynamic derivatives depends on the type of variables which are kept constant under the given experimental conditions [38]. When the experimental conditions are such that only field variables are fixed, with no fixed density variables, then the derivative of a field with respect to a density, such as $(\partial A/\partial \xi)_{eq}$, will approach zero as $|T - T_{cr}|^x$ with a "strong" critical index x of order unity. However, if among the fixed variables there is one density variable, the critical index x is "weak" (of order 0.1). Finally, if two or more density variables are fixed, then $x = 0$, and the derivatives will be continuous. Applying the Griffiths-Wheeler rule to chemical reactions near the critical points, one has to calculate the number of fixed densities. The latter will include all inert components which are not participating in any chemical reaction.

From the Griffiths-Wheeler division of the thermodynamic variables into "field" and "density", it follows that the approach to a critical point is path-dependent, as is shown in Fig. 2.3 for the critical point of a one-component fluid [39].

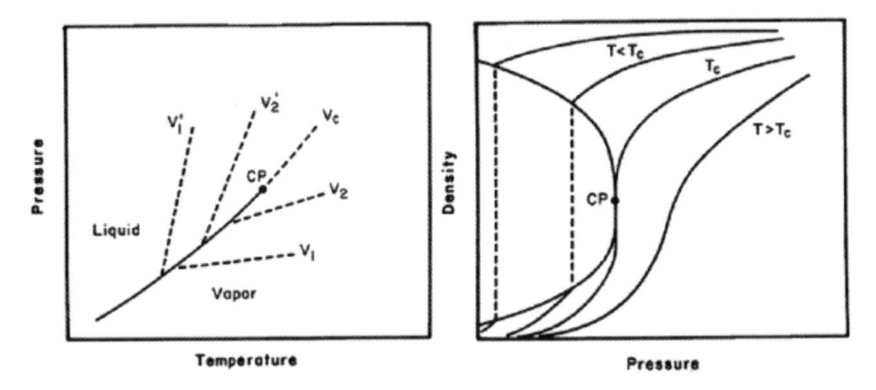

Fig. 2.3 Schematic pressure-temperature and density-pressure diagram for a pure fluid. Reproduced from Ref. [39] with permission, copyright (1987), International Union of Pure and Applied Chemistry.

If one approaches the critical point along the critical isotherm, the isothermal compressibility, $(\partial \rho/\partial p)_{T=T_C}$, diverges with the strong critical index $x = 1.24$. However, approaching the critical point along the critical isochore results in different behavior. Neither pressure or temperature changes rapidly upon approaching the critical point, and the derivative $(\partial p/\partial T)_{v=vC}$ remains finite at the critical point. By contrast, the specific heat at constant volume, $C_v \equiv T\,(\partial S/\partial T)_{v=v_{cr}}$ (derivative of a "density" with respect to "field" with one density held constant), is characterized by the weak critical index $x = 0.11$. Similar arguments can be used for the analysis of the thermodynamic variables near the critical points in many-component systems.

The preceding thermodynamic analysis is not complete, for the following reason. Only homogeneous changes of the extent of reaction ξ are allowed by Eq. (2.9). However, if a system is large enough, the k -dependent changes connected with sound, heat or diffusion modes become important [40]. Indeed, the relaxation rates of the latter processes will become smaller for some small wavenumber k than the fixed k -independent rate of the chemical reaction. The general mean-field approach, which goes back to [35], is the straightforward generalization of (2.9). On the left-hand side of the equation, instead of ξ, appears the column matrix of the density-type variables x, and instead of A, their conjugate fields appear (pressure-density, chemical potentials-concentrations, temperature-entropy, etc.)

$$i\omega x\left(k,\omega\right)=L\left(k\right)X\left(k,\omega\right). \tag{2.21}$$

The field variables X are related, in turn, to the density variables x by the susceptibility matrix χ^{-1} by $X\left(k,\omega\right)=\chi\left(k\right)^{-1}x\left(k,\omega\right)$ (the constituent equation), leading to

$$i\omega x\left(k,\omega\right)=L\left(k\right)\chi\left(k\right)^{-1}x\left(k,\omega\right)\equiv Mx\left(k,\omega\right). \tag{2.22}$$

The matrix $M=L\left(k\right)\chi\left(k\right)^{-1}$ is the hydrodynamic matrix, which we will analyze in detail when we consider the hydrodynamic equations for a reactive system.

As emphasized in [40], Eq. (2.21) shows that the relaxation of the extent of reaction ξ arises not only through the chemical processes (with characteristic time $\tau\equiv\left[L\left(\partial A/\partial\xi\right)_{eq}\right]^{-1}$ away from the critical point), but also through diffusion and heat conductivity (with characteristic times $\left(Dk^2\right)^{-1}$ and $\left(\lambda k^2/\rho C_p\right)^{-1}$, respectively). Therefore, one must compare the relaxation times or characteristic lengths (inverse wavenumbers) in order to determine which quantity should be considered constant. As we have seen, this is of crucial importance for the analysis of slowing-down processes.

Let us define two characteristic lengths by $k_C^{-2}=D\tau$ and $k_H^{-2}=\lambda\tau/\rho C_p$, where, generally, $k_H^{-1}>>k_C^{-1}$. Values of the wavenumbers k fall into three intervals: $k<k_H$, $k_H<k<k_C$, and $k>k_C$. In the first interval, the chemical reaction is more rapid than both heat conductivity and diffusion, whereas in the third interval, diffusion dominates the chemical relaxation. Only for processes with wavenumbers located in the "window" between k_H and k_C is the relaxation dominated by chemistry. For typical values of the heat conductivity, $\lambda/\rho C_p=0.1\ \mathrm{cm}^2/\mathrm{s}$, the diffusion constant

$D = 10^{-5}$ cm^2/s, and for a reaction rate $\tau^{-1} = 120$ Hz, we have $k_H = 10^3$ cm^{-1} and $k_C = 10^5$ cm^{-1}. Therefore, light scattering experiments with $k \approx 10^4$ cm^{-1} at different temperatures in the vicinity of the critical temperature provides evidence for the slowing-down of chemical reactions near the critical points.

Note that the "slowing-down" of the chemical reaction does not imply that the forward or backward reaction is slowed down. It is the measured rate, which is the net difference between the forward and backward reactions, which is affected by criticality. In fact, condition (2.19), which is really $(\partial A/\partial \xi)_{T,p} \approx 0$, means that the system becomes indifferent to changes in the species concentration. In equilibrium, when $A = 0$, the reaction is balanced and the measured rate is zero. Usually, a change of ξ from ξ_{eq} builds up an affinity $A \neq 0$ which acts as a driving force to restore the equilibrium. However, due to the thermodynamic properties in the critical region, a change in ξ does not create a restoring force (i.e., affinity), and the reaction continues to be balanced even though $\xi \neq \xi_{eq}$. That is, the rate continues to be zero. In other words, the physical explanation of the slowing-down of the chemical reaction is the same as the explanation of diffusion near the critical points of nonreactive systems, where the vanishing of diffusion does not mean that the individual particles are slowing-down. The only difference is that the chemical perturbation is homogeneous, in contrast to the inhomogeneous diffusion changes.

All the above considerations hold for a two-component mixture with a single chemical reaction. If the system has m reactants and n products, then there are $m+n-2$ conserved variables. If all $m+n$ species in the system take part in the reaction, then the reaction can slow-down strongly for $m+n < 3$, weakly for $m + n = 3$, and not at all for $m + n > 3$. Therefore, for more complicated systems than a binary mixture, strong critical slowing-down is not likely to occur in the usual experimental situation. The additional factor which has to be taken into account in analyzing an experiment, is the path of approach to the critical point (at some fixed parameter(s), along the coexistence curve, etc.), since the behavior of the thermodynamic derivatives is path-dependent.

2.4 Hydrodynamic equations of reactive binary mixture: piston effect

The slowing-down of the dynamic processes near the liquid-gas critical points, considered in the previous section, is the hallmark of critical phenomena. It is precisely this fact which makes it so difficult to obtain reliable equilibrium experimental data near the critical points.

However, the very interesting phenomenon of the speeding-up of dynamic processes in critical fluids at constant volume, as opposed to their slowing-down, was predicted theoretically [41], and afterwards found experimentally [42]. The usual experimental set-up is a fluid-filled closed cell. After changing the temperature at the bottom or around the cell, one waits for the establishment of thermal equilibrium inside the cell. The slow diffusive heat propagation is responsible for the critical slowing-down. However, fast thermal equilibrium is established through the thermo-acoustic effect [41], where the temperature change at the surface induces acoustic waves which lead to a fast change of the pressure, and hence, of the temperature everywhere in the fluid. This is also called the "piston effect", since the thermal boundary layer generated near the heated bottom of the cell produces, like a piston, the pressure on the fluid restricted to a constant volume.

The speeding-up of an interface reaction, which is immediately evident from the piston effect, has been explained in [43]. There are also other explanations for the speeding-up of chemical reactions near critical points based on specific properties, such as appearance of the soft modes in the vibrational spectrum of solids [44], reduction of the H-bond strength in water [45], and the solid-solvent clustering mechanism [46]. We shall consider some special examples.

2.4.1 *Heterogeneous reactions in near-critical systems*

The interface reaction that takes place in a supercritical phase may be cited as an illustration of an immediate influence of the piston effect on the flux of matter at the interface, leading to the speeding-up of the chemical reaction [43], [47]. Such speeding-up manifests itself in the strong corrosion observed in supercritical fluid containers [48]. Two factors — the sensitivity of solubility to the small changes of pressure, induced by the piston effect, and the supercritical hydrodynamics — affect the absorption-desorption reaction taking place between a solid interface and an active component in a dilute binary supercritical mixture. The method of matched asymptotic expansion has been used [43] for the solution of the simplified system of one-dimensional hydrodynamic equations with a van der Waals equation of state and linear mixing rules. The asymptotic analysis shows the coupling between the piston effect and an interface reaction, which — for some initial conditions — leads to a strong increase of the flux of matter and the intensification of a reaction.

More detailed analysis has been performed [47]. The numerical solution of the full hydrodynamic equations based on the finite volume approximation has been carried out for a system involving a small amount of naphthalene dissolved in supercritical CO_2 located between two infinite solid plates made of pure solid naphthalene and able to absorb or desorb naphthalene from the fluid. The fluid is initially at rest. The temperature of one of the plates is then gradually increased while the other plate is kept at the initial temperature. Due to the small amount of naphthalene in solution, for numerical simulation one can use the parameters of pure CO_2. The results of the analysis show that there are three mechanisms for the increase of the mass fractions of the naphthalene w at the solid interface:

1. The critical behavior of the solute solubility through the large values of $(\partial w/\partial p)_T$, and (at the heated plate) of $(\partial w/\partial T)_p$.

2. The strong compressibility of the supercritical fluid through the derivative $(\partial w/\partial p)_T$, and the piston effect that results through the pressure variation.

3. The strong dilatability expressed by the large derivative $(\partial w/\partial T)_p$ and through the density variation.

The piston effect, coupled with the critical behavior of the solute solubility with respect to pressure, is the leading mechanism governing the mass fraction of naphthalene at the isothermal plate [47]. Similar simulations have been performed with cooling, rather than heating one of the plates, and a similar phenomenon was observed: the strong decrease of the pressure caused by the piston effect leads to a strong decrease of the mass fraction of naphthalene at the isothermal plate.

In the next section, we carry out a phenomenological analysis based on the full hydrodynamic equations in a reactive mixture [49], [50].

2.4.2 *Dynamics of chemical reactions*

There are two different phenomenological ways to describe chemical reactions: the hydrodynamic approach, used by physicists and which we use here, and the rate equations, used mostly by chemists.

We consider an isomerization reaction between two species which is described by the equation $A \rightleftarrows B$. This simple example of a chemical reaction can be easily generalized to more complex reactions. The question arises as to the choice of the thermodynamic variables which characterize the binary mixture. A one-component thermodynamic system is characterized by two parameters, such as pressure and temperature. For a binary mixture, one has to add one additional parameter, such as the concentration ξ of one of the components, which will also describe a chemical reaction. Indeed, for binary mixtures, the chemical potential μ, defined [4] as $\mu = \mu_1/m_1 - \mu_2/m_2$, where m_1 and m_2 are the molecular masses of the components, is proportional to the affinity $A = \mu_1\nu_1 - \mu_2\nu_2$ of the chemical reaction since the stoichiometric coefficients ν_1, ν_2 are inversely proportional to the molecular masses of the species. Thus, one can choose the mass fraction of one component as the progressive variable ξ of a chemical reaction. The dynamic behavior of a reactive binary mixture, described by the hydrodynamic equations, will contain a fourth variable, which for an isotropic system can be chosen as $div\,(v)$.

As discussed in the previous section, near equilibrium in the linear approximation, one obtains

$$\frac{d\xi}{dt} = -L_0 A = -L_0 \left(\frac{dA}{d\xi}\right)_{eq} \left(\xi - \xi_{eq}\right). \tag{2.23}$$

The subscript eq means that the thermodynamic derivative is calculated in equilibrium, and L_0 is the Onsager coefficient. The quantity $\left[L_o \left(\partial A/\partial\xi\right)_{eq}\right]^{-1}$ is the characteristic time of the chemical reaction described by the phenomenological equation (2.23). In real system, however, the chemical reaction occurs in some media, and the "chemical" mode interacts with other hydrodynamic modes, such as diffusive, viscous and heat conductive modes, i.e., one has to consider the entire system of hydrodynamic equations. This requires calculating the "renormalized" Onsager coefficient L due to the interaction with the hydrodynamic modes ("mode coupling"). In equilibrium statistical mechanics, there exists a general procedure for calculating the thermodynamic quantities, which goes back to Gibbs. No such systematic procedure exists in non-equilibrium statistical mechanics. However, for non-equilibrium states that are not far from equilibrium, some general formulae do exist which connect the kinetic coefficients with long-wavelength limits of the correlation functions. The Onsager coefficient plays a role analogous to that of transport coefficients in hydrodynamics, and both are described by the Green-Mori-Zwanzig relations [51]–[53]. The "chemical" mode is characterized by the local interactions whereas the pure hydrodynamic modes depend upon spatial gradients. In a real fluid, there are couplings between different modes, and the Onsager coefficient L is defined for small wavenumbers k, as the correlation function of the changes of the concentration $(\partial\xi/\partial t)_{chem}$ due to the chemical reaction [54],

$$L = \frac{1}{\kappa_B T_0} \int_0^\infty dt \left\langle \left(\frac{\partial\xi}{\partial t}\right)_{chem} (k,t) \left(\frac{\partial\xi}{\partial t}\right)_{chem} (-k,0) \right\rangle. \tag{2.24}$$

Equation (2.24) provides the connection between the phenomenological Onsager coefficient and the microscopic time correlation function. First of all, let us check that in the absence of the hydrodynamic modes, when the affinity A depends only on the concentration ξ rather than on all thermodynamic variables, the renormalized Onsager coefficient L, defined in (2.24), reduces to the "bare" Onsager coefficient L_0. Indeed, the solution of Eq. (2.23) has the form

$$\xi - \xi_{eq} = \exp\left[-L_0 \left(\frac{dA}{d\xi}\right)_{eq} t\right]. \tag{2.25}$$

Substituting (2.23) and (2.25) into (2.24) yields

$$L = L_0^2 \left(\frac{dA}{d\xi}\right)_{eq}^2 \frac{1}{\kappa_B T_0} \int_0^\infty dt \exp\left[-L_0 \left(\frac{dA}{d\xi}\right)_{eq} t\right] \left\langle \xi(k,0)\, \xi(-k,0) \right\rangle. \tag{2.26}$$

Performing the integration over t in (2.26), and taking into account [4] that the equilibrium correlation function $\langle \xi(k,0)\, \xi(-k,0) \rangle$ equals $\kappa_B T_0 / (dA/d\xi)_{eq}$, one immediately obtains that $L = L_0$, as requires for a purely chemical process. In the following section, we will consider the influence of the hydrodynamic modes on the renormalized Onsager coefficient.

2.4.3 *Relaxation time of reactions*

Consider the isomerization reaction with c_1 and c_2 being the mass concentrations of substances A and B, and γ_{ij} are the transition rates from component i to component j. The change of the concentration of component A caused by gain and loss to component B, is described by

$$\frac{dc_1}{dt} = \gamma_{21} c_2 - \gamma_{12} c_1 = \gamma_{21} - (\gamma_{12} + \gamma_{21})\, c_1. \tag{2.27}$$

For a closed system the total number of particles is constant, $c_1 + c_2 = 1$, which has been used in Eq. (2.27). This equation contains the characteristic time of the reaction r_0 which depends on the rate constants γ_{12} and γ_{21} at equilibrium,

$$r_0 = \frac{1}{\gamma_{12} + \gamma_{21}}. \tag{2.28}$$

At equilibrium, $dc_1/dt = 0$, and

$$\gamma_{12} c_1^0 = \gamma_{21} c_2^0. \tag{2.29}$$

The kinetic processes lead to the renormalization of this characteristic time, which is described by Eq. (2.24) with a slight change in the constant coefficient,

$$r = \frac{1}{c_1^0 c_2^0} \int_0^\infty dt \left\langle \left(\frac{\partial c}{\partial t} \right)_{chem} (k,t) \left(\frac{\partial c}{\partial t} \right)_{chem} (-k,0) \right\rangle. \tag{2.30}$$

If a system is open, the particles are able to enter and leave. Then, the concentrations depend both on chemical transformations and on diffusion, and their time dependence is described by the following equations [55],

$$\begin{aligned} \partial c_1/\partial t &= D_1 \nabla^2 c_1 - \gamma_{12} c_1 + \gamma_{21} c_2, \\ \partial c_2/\partial t &= D_2 \nabla^2 c_1 - \gamma_{21} c_2 + \gamma_{12} c_1. \end{aligned} \tag{2.31}$$

We use the Fourier and Laplace transforms in space and time, respectively,

$$c_i(k,s) = \iint d\mathbf{r} dt \exp(-i\mathbf{kr}) \exp(-st)\, c_i(\mathbf{r},t). \tag{2.32}$$

The solutions of Eqs. (2.31) are

$$\begin{aligned} c_1(k,s) &= \frac{\left(s + k^2 D_1 + \gamma_{21}\right) c_1(k,0) + \gamma_{21} c_2(k,0)}{\Delta}, \\ c_2(k,s) &= \frac{\left(s + k^2 D_2 + \gamma_{12}\right) c_2(k,0) + \gamma_{12} c_1(k,0)}{\Delta}, \end{aligned} \tag{2.33}$$

where

$$\begin{aligned} \Delta = s^2 &+ s\left[(D_1 + D_2) k^2 + \gamma_{21} + \gamma_{12}\right] \\ &+ k^2 (D_1 \gamma_{12} + D_2 \gamma_{21}) + k^4 D_1 D_2. \end{aligned} \tag{2.34}$$

Multiplying Eqs. (2.33) by $c_1(-k,0)$ and $c_2(-k,0)$ and averaging, one obtains the expressions for the matrix elements $\langle c_i(k,s)\, c_j(-k,0) \rangle$, for

$i, j = 1, 2$, as functions of the equilibrium averages $\langle c_i(k,0)\, c_j(-k,0)\rangle$. For special systems, such as structurally similar isomers, for which $D_1 = D_2 \equiv D$, one gets

$$\langle c_1(k,s)\, c_1(-k,0)\rangle = \frac{\left[s + k^2 D + \gamma_{21}\right] \langle c_1(k,0)\, c_1(-k,0)\rangle + \gamma_{21} \langle c_2(k,0)\, c_1(-k,0)\rangle}{s^2 + s\left(\gamma_+ + 2k^2 D\right) + k^2 \gamma_+ D + k^4 D^2}, \quad (2.35)$$

$$\langle c_2(k,s)\, c_2(-k,0)\rangle = \frac{\left[s + k^2 D + \gamma_{12}\right] \langle c_2(k,0)\, c_2(-k,0)\rangle + \gamma_{12} \langle c_1(k,0)\, c_2(-k,0)\rangle}{s^2 + s\left(\gamma_+ + 2k^2 D\right) + k^2 \gamma_+ D + k^4 D^2}, \quad (2.36)$$

$$\langle c_1(k,s)\, c_2(-k,0)\rangle = \frac{\left[s + k^2 D + \gamma_{21}\right] \langle c_1(k,0)\, c_2(-k,0)\rangle + \gamma_{21} \langle c_2(k,0)\, c_2(-k,0)\rangle}{s^2 + s\left(\gamma_+ + 2k^2 D\right) + k^2 \gamma_+ D + k^4 D^2}, \quad (2.37)$$

$$\langle c_2(k,s)\, c_1(-k,0)\rangle = \frac{\left[s + k^2 D + \gamma_{12}\right] \langle c_2(k,0)\, c_1(-k,0)\rangle + \gamma_{12} \langle c_1(k,0)\, c_1(-k,0)\rangle}{s^2 + s\left(\gamma_+ + 2k^2 D\right) + k^2 \gamma_+ D + k^4 D^2}, \quad (2.38)$$

where $\gamma_+ = \gamma_{21} + \gamma_{12}$.

Equations (2.35)–(2.38) are simplified when both components are taken to be ideal gases. Then, $\langle c_i(k,0)\, c_j(-k,0)\rangle = c_j^0 \delta_{ij}$, so that

$$\langle c_1(k,s)\, c_1(-k,0)\rangle = c_1^0 \frac{s + k^2 D + \gamma_{21}}{s^2 + s\left(\gamma_{21} + \gamma_{12} + 2k^2 D\right) + k^2 \gamma_+ D + k^4 D^2}, \quad (2.39)$$

$$\langle c_2(k,s)\, c_2(-k,0)\rangle = c_2^0 \frac{\left[s + k^2 D + \gamma_{12}\right]}{s^2 + s\left(\gamma_{21} + \gamma_{12} + 2k^2 D\right) + k^2 \gamma_+ D + k^4 D^2}, \quad (2.40)$$

$$\langle c_1(k,s)\, c_2(-k,0)\rangle = c_2^0 \frac{\gamma_{21}}{s^2 + s\left(\gamma_{21} + \gamma_{12} + 2k^2 D\right) + k^2 \gamma_+ D + k^4 D^2}, \quad (2.41)$$

$$\langle c_2(k,s)\, c_1(-k,0)\rangle = c_1^0 \frac{\gamma_{12}}{s^2 + s\left(\gamma_{21} + \gamma_{12} + 2k^2 D\right) + k^2 \gamma_+ D + k^4 D^2}. \quad (2.42)$$

Substituting into (2.30) the expression $(\partial c/dt)_{chem} = \gamma_{21}c_2 - \gamma_{12}c_1$, one obtains

$$r = \frac{1}{c_1^0 c_2^0}\int_0^\infty dt \int_0^\infty ds \, \exp(st)\left\{\gamma_{12}^2 \langle c_1(k,s)\,c_1(-k,0)\rangle\right.$$
$$+\gamma_{21}^2 \langle c_2(k,s)\,c_2(-k,0)\rangle$$
$$\left.-\gamma_{12}\gamma_{21}\left[\langle c_1(k,s)\,c_2(-k,0)\rangle + \langle c_2(k,s)\,c_1(-k,0)\rangle\right]\right\} \tag{2.43}$$

which takes the following form after using Eqs. (2.39)–(2.42),

$$r = \frac{\left(\gamma_{12}^2 c_1^0 + \gamma_{21}^2 c_2^0\right)}{c_1^0 c_2^0}$$
$$\times \int_0^\infty dt \int_0^\infty ds \frac{\left(s + k^2 D\right)\exp(st)}{s^2 + s\left(\gamma_+ + 2k^2 D\right) + k^2\gamma_+ D + k^4 D^2}. \tag{2.44}$$

To perform the inverse Laplace transformation on Eq. (2.44), one needs the roots of the denominator,

$$s_1 = -k^2 D; \qquad s_2 = -k^2 D - \gamma_+. \tag{2.45}$$

Integrating over s yields

$$r = \frac{\gamma_{12}^2 c_1^0 + \gamma_{21}^2 c_2^0}{c_1^0 c_2^0\left(\gamma_{12} + \gamma_{21}\right)}\int_0^\infty \exp\left[-\left(k^2 D + \gamma_+\right)t\right]dt$$
$$= \frac{\gamma_{12}^2 c_1^0 + \gamma_{21}^2 c_2^0}{c_1^0 c_2^0\left(\gamma_{12} + \gamma_{21}\right)\left(k^2 D + \gamma_{21} + \gamma_{12}\right)} \tag{2.46}$$

which can be rewritten by using Eq. (2.29),

$$r = \frac{1}{\left(k^2 D + \gamma_{21} + \gamma_{12}\right)}. \tag{2.47}$$

For $D = 0$, this solution reduces to Eq. (2.28), as required.

2.4.4 *Hydrodynamic equations of a reactive binary mixtures*

The hydrodynamic equations are the conservation laws of mass, momentum, concentration and energy. They are the continuity equation,

$$\frac{d\rho}{dt} = -div\left(\rho v\right), \tag{2.48}$$

the Navier-Stokes equation,

$$\rho \frac{dv_i}{dt} = -\frac{\partial}{\partial x_j} \sigma_{ij}, \tag{2.49}$$

the conservation of one of the components,

$$\frac{d\xi}{dt} = -div\left(J_\xi\right), \tag{2.50}$$

and the conservation of energy,

$$\frac{dQ}{dt} = \rho T \frac{dS}{dt} = -div\left(J_\epsilon\right). \tag{2.51}$$

Just as ρv in Eq. (2.48) represents the mass flux, the terms σ_{ij}, J_ξ, and J_ϵ in (2.49)–(2.51) describe the stress tensor, fluxes of mass of one of the components and energy, respectively. The phenomenological forms of these quantities, which include both hydrodynamic and chemical flows, are based on the symmetry requirements for vectors J, v and scalars ρ, ξ, S and the chemical affinity A.

The well-known expressions for hydrodynamic fluxes are [4]

$$\left(\sigma_{ij}\right)_{hydr} = \delta_{ij} p + \eta \left[\frac{\partial v_i}{\partial x_j} + \frac{\partial v_j}{\partial x_i} - \frac{2}{3}\delta_{ij} div\left(v\right)\right] + \delta_{ij}\zeta div\left(v\right), \tag{2.52}$$

$$\left(J_\xi\right)_{hydr} = -Dgrad\left(\xi\right) - D\left[\frac{k_T}{T} grad\left(T\right) + \frac{k_p}{p} grad\left(p\right)\right], \tag{2.53}$$

$$\left(J_\epsilon\right)_{hydr} = -\lambda grad\left(T\right) - \rho D \left(\frac{\partial A}{\partial \xi}\right)_{p,T} \left[k_T grad\left(\xi\right) + \frac{k_p}{p} grad\left(p\right)\right]. \tag{2.54}$$

The variables appearing in (2.48)–(2.54) have obvious meaning. The coefficient k_T defines the Dufour and Soret cross-effects, whereas k_p/p is a purely thermodynamic quantity defined by

$$\frac{k_p}{p} = \frac{\left(\partial A/\partial p\right)_{T,\xi}}{\left(\partial A/\partial \xi\right)_{p,T}}. \tag{2.55}$$

For small deviations from chemical equilibrium, the constitutive equation takes the form

$$\left(\frac{d\xi}{dt}\right)_{chem} = -L_0 A. \tag{2.56}$$

The heat change in the chemical reaction is given by

$$\lambda = \frac{\Delta h}{C} \tag{2.57}$$

where C is the heat capacity, and Δh is the change of enthalpy per unit mass of reaction. The heat flux caused by the chemical reaction is

$$\left(\frac{dQ}{dt}\right)_{chem} = \lambda L_0 A. \tag{2.58}$$

Since Δh is proportional to A, the λLA term in Eq. (2.58) is proportional to A^2, and can therefore be neglected for small deviations from equilibrium, as is assumed when using the hydrodynamic theory.

The phenomenological relations (2.56) and (2.52)–(2.54) are based on symmetry considerations which — in the presence of a chemical reaction — allow the appearance of an additional visco-reactive cross effects, described by the L_1 terms. This leads to the following generalization of Eqs. (2.56) and (2.52),

$$\left(\frac{d\xi}{dt}\right)_{chem} = -L_0 A - L_1 div\,(v) \tag{2.59}$$

$$(\sigma_{ij})_{chem} = \delta_{ij} L_1 A. \tag{2.60}$$

Inserting (2.53), (2.59), and (2.60) into (2.49)–(2.51), yields four linearized hydrodynamic equations,

$$\frac{\partial \rho}{\partial t} = -\rho_0 div\,(v) \tag{2.61}$$

$$\rho_0 \frac{\partial v}{\partial t} = -grad\,(p) + \eta \nabla^2 v + \left(\frac{\eta}{3} + \zeta\right) grad\,(div\,(v)) + \rho_0 L_1 grad\,(A) \tag{2.62}$$

$$\frac{\partial \xi}{\partial t} = D\nabla^2 \xi + D\left(\frac{k_T}{T}\nabla^2 T + \frac{k_p}{p}\nabla^2 p\right) - L_0 A - L_1 div\,(v) \tag{2.63}$$

$$\rho T \frac{\partial S}{\partial t} = \lambda \nabla^2 T + \rho D \left(\frac{\partial A}{\partial \xi}\right)_{p,T} \left(k_T \nabla^2 \xi + \frac{k_p}{p}\nabla^2 p\right) \tag{2.64}$$

where ρ, p, T and ξ are the deviations of density, pressure, temperature and the mass fraction of one of the components from their equilibrium values denoted by the subscript zero. The transport coefficients in Eqs. (2.61)–(2.64) are the shear viscosity η, the bulk viscosity ς, the heat-conductivity λ, the diffusion coefficient D, the thermal-diffusion coefficient k_T, and the baro-diffusion coefficient k_p.

The rigorous treatment of the correlation functions, which, according to Eq. (2.24), define the renormalized Onsager coefficient L, requires that allowance be made for the symmetry properties of the variables entering the hydrodynamic equations (2.61)–(2.64). The quantity $(\partial\xi/\partial t)_{chem}$ and the variable $div\,(v)$ are odd functions under time reversal and even functions under parity [36], while the other variables are either even or odd under both symmetry operations. However, these arguments are not relevant in the following, since we use simplified models.

Using Eq. (2.61), one can omit $div\,(v)$ from Eqs. (2.62)–(2.63), which gives

$$-\frac{\partial^2\rho}{\partial t^2} = -\nabla^2 p - \frac{1}{\rho_0}\left(\frac{4}{3}\eta + \zeta\right)\nabla^2\frac{\partial\rho}{\partial t} + \rho_0 L_1 \nabla^2 A \tag{2.65}$$

$$\frac{\partial\xi}{\partial t} = D\nabla^2\xi + D\left(\frac{k_T}{T}\nabla^2 T + \frac{k_p}{p}\nabla^2 p\right) - L_0 A + \frac{L_1}{\rho_0}\frac{\partial\rho}{\partial t}. \tag{2.66}$$

To solve Eqs. (2.64)–(2.66), we express A, T and p in terms of the variables ρ, ξ, S. In equilibrium, $A = 0$. For small deviations from equilibrium,

$$A = \left(\frac{\partial A}{\partial\rho}\right)_{\xi,S}\rho + \left(\frac{\partial A}{\partial S}\right)_{\rho,\xi} S + \left(\frac{\partial A}{\partial\xi}\right)_{S,\rho}\xi. \tag{2.67}$$

Our analysis is based on the assumption of local equilibrium, i.e.,

$$\begin{aligned} T &= (\partial T/\partial\rho)_{\xi,S}\,\rho + (\partial T/\partial S)_{\rho,\xi}\,S + (\partial T/\partial\xi)_{S,\rho}\,\xi, \\ p &= (\partial p/\partial\rho)_{\xi,S}\,\rho + (\partial p/\partial S)_{\rho,\xi}\,S + (\partial p/\partial\xi)_{S,\rho}\,\xi. \end{aligned} \tag{2.68}$$

Inserting (2.67) and (2.68) into (2.64)–(2.66) yields [56] a quite cumbersome system of three differential equations in three variables ρ, ξ, S. The Fourier and Laplace transforms in space and time, respectively, yield the following solution of these equations,

$$\xi\,(k,s) = \frac{f_1\,(k,s)\,\xi\,(k,0) + f_2\,(k,s)\,\rho\,(k,0) + f_3\,(k,s)\,S\,(k,0)}{\Delta}, \tag{2.69}$$

$$\rho\,(k,s) = \frac{f_4\,(k,s)\,\xi\,(k,0) + f_5\,(k,s)\,\rho\,(k,0) + f_6\,(k,s)\,S\,(k,0)}{\Delta}, \tag{2.70}$$

$$S\,(k,s) = \frac{f_7\,(k,s)\,\xi\,(k,0) + f_8\,(k,s)\,\rho\,(k,0) + f_9\,(k,s)\,S\,(k,0)}{\Delta}, \tag{2.71}$$

where the determinant of the coefficients in the original equations, Δ, and the functions f_i $(i = 1, \ldots, 9)$ are cumbersome combinations of the thermodynamic variables and transport coefficients, which can be found in [56].

The complicated hydrodynamic equations can be simplified if one neglects the higher terms in k^2. Performing the Fourier-Laplace transforms of Eqs. (2.65) and (2.66), written in the variables ρ and ξ, one gets

$$\begin{aligned} & K_{1,\rho}\rho(k,s) + K_{1,\xi}\xi(k,s) \\ & = \left[s - \frac{1}{\rho_0}\left(\frac{4}{3}\eta + \zeta\right)k^2\right]\rho(k,0) + \partial\rho(k,0)/\partial t, \\ & K_{2,\rho}\rho(k,s) + K_{2,\xi}\xi(k,s) = \xi(k,0) - (L_1/\rho_0)\,\rho(k,0), \end{aligned} \tag{2.72}$$

where one can set $\partial\rho(k,0)/\partial t = 0$, and

$$K_{1,\rho} = s^2 + k^2\left[\left(\frac{\partial p}{\partial \rho}\right)_{\xi,S} + \left(\frac{4}{3}\gamma + \zeta\right)\frac{s}{\rho_0} + \rho_0 L_1\left(\frac{\partial A}{\partial \rho}\right)_{\xi,S}\right], \tag{2.73}$$

$$K_{1,\xi} = k^2\left[\left(\frac{\partial p}{\partial \xi}\right)_{\rho,S} + \rho_0 L_1\left(\frac{\partial A}{\partial \xi}\right)_{\rho,S}\right], \tag{2.74}$$

$$K_{2,\rho} = L_0\left(\frac{\partial A}{\partial \rho}\right)_{\xi,S} + Dk^2\left(\frac{\partial \xi}{\partial \rho}\right)_{\xi,S} - \frac{sL_1}{\rho_0}, \tag{2.75}$$

$$K_{2,\xi} = s + k^2 D + L_0\left(\frac{\partial A}{\partial \xi}\right)_{\rho,S}. \tag{2.76}$$

The solutions of Eqs. (2.72) are

$$\begin{aligned} & \rho(k,s) \\ & = \frac{\left\{K_{2,\xi}\left[s - \frac{1}{\rho_0}\left(\frac{4}{3}\eta + \zeta\right)k^2\right] - K_{1,\xi}\frac{L_1}{\rho_0}\right\}\rho(k,0) - K_{1,\xi}\xi(k,0)}{K_{1,\rho}K_{2,\xi} - K_{2,\rho}K_{1,\xi}} \end{aligned} \tag{2.77}$$

$$\begin{aligned} & \xi(k,s) \\ & = \frac{K_{1,\rho}\left[\xi(k,0) - \frac{L_1}{\rho_0}\rho(k,0)\right] - K_{2,\rho}\left[s - \frac{1}{\rho_0}\left(\frac{4}{3}\eta + \zeta\right)k^2\right]\rho(k,0)}{K_{1,\rho}K_{2,\xi} - K_{2,\rho}K_{1,\xi}}. \end{aligned} \tag{2.78}$$

To lowest order in k^2, Eq. (2.77) takes the following form

$$\rho(k,s) = s\rho(k,0)\left\{s^2 + \left(sk^2/\rho_0\right)\left(\frac{4}{3}\gamma + \zeta\right)\right.$$

$$+ \left[(\partial p/\partial\rho)_{\xi,S} + \rho_0 L_1 (\partial A/\partial\rho)_{\xi,S}\right] k^2$$

$$+ \left[(\partial p/\partial\xi)_{\xi,S} + \rho_0 L_1 \partial A/\partial\xi_{\xi,S}\right]\left[sL_1/\rho_0 - L_0 (\partial A/\partial\xi)_{\rho,S}\right]$$

$$\left.\times \left[s + L_0 (\partial A/\partial\xi)_{\rho,S}\right]^{-1} k^2\right\}^{-1}$$

$$\approx s\rho(k,0)\left\{\left[s + \left(k^2/2\rho_0\right)\left(\frac{4}{3}\gamma + \zeta\right)\right]^2\right.$$

$$\left. + \left[(\partial p/\partial\rho)_{\xi,S} + \rho_0 L_1 (\partial A/\partial\rho)_{\xi,S}\right] k^2\right\}^{-1}. \quad (2.79)$$

The inverse Laplace transform yields

$$\rho(k,t) = \rho(k,0)\exp\left[-\frac{1}{2\rho_0}\left(\frac{4}{3}\gamma + \zeta\right)k^2 t\right]$$

$$\times \cos\left\{\left[\left(\frac{\partial p}{\partial\rho}\right)_{\xi,S} + L_1\left(\frac{\partial A}{\partial\rho}\right)_{\xi,S}\right]^{1/2} kt\right\}. \quad (2.80)$$

Analogously, to lowest order in k^2, one gets from Eq. (2.78),

$$\xi(k,t)$$

$$= \left[\xi(k,0) - \frac{L_1}{\rho_0}\rho(k,0)\right]\exp\left[-\left(Dk^2 + L_0\left(\frac{\partial A}{\partial\xi}\right)_{\rho,S}\right)t\right]. \quad (2.81)$$

According to Eq. (2.24), the rate of the chemical reaction is defined by the change of concentration $(\partial\xi/\partial t)_{chem}$ due to the chemical reaction. In the case under consideration, this change is given by the term $-L_0 A$ in Eq. (2.66). Using ξ and ρ as independent variables, one gets

$$L = \left(L_0^2/k_B T_0\right)\int_0^\infty dt\,\langle A(k,t)\,A(-k,0)\rangle$$

$$= \left(L_0^2/k_B T_0\right)(\partial A/\partial\xi)^2_{\rho,S}\int_0^\infty dt[\langle\xi(k,t)\,\xi(-k,0)\rangle$$

$$+ (\partial\xi/\partial\rho)^2_{A,S}\langle\rho(k,t)\,\rho(-k,0)\rangle$$

$$- (\partial\xi/\partial\rho)_{A,S}(\langle\xi(k,t)\,\rho(-k,0)\rangle + \langle\rho(k,t)\,\xi(-k,0)\rangle)]. \quad (2.82)$$

The correlators in Eq. (2.82) can easily be found from Eqs. (2.80) and (2.81) as functions of equilibrium correlators

$$\begin{aligned}
\langle \rho(k,0)\,\rho(-k,0)\rangle &= \kappa T_0 \rho_0^2 \left(\partial p/\partial \rho\right)_{\xi,S}^{-1}, \\
\langle \xi(k,0)\,\xi(-k,0)\rangle &= \kappa T_0 \left(\partial A/\partial \xi\right)_{\rho,S}^{-1}, \\
\langle \xi(k,0)\,\rho(-k,0)\rangle &= -\kappa T_0 \left(\partial \rho/\partial A\right)_{\xi,S}.
\end{aligned} \tag{2.83}$$

Finally, one obtains

$$\begin{aligned}
L = L_0^2 \left(\partial A/\partial \xi\right)_{p,S}^2 &\Bigg\{ \Bigg[\frac{L_1}{\rho_0}\left[Dk^2 + L_0 \left(\partial A/\partial \xi\right)_{\rho,S}\right]^{-1} \\
&+ \left(\frac{\partial \xi}{\partial \rho}\right)_{A,S} \frac{2\left(4\eta/3+\zeta\right)\rho_0}{\left(4\eta/3+\zeta\right)^2 k^2 + 4\left[\left(\partial p/\partial \rho\right)_{\xi,S} + \rho_0 L_1 \left(\partial A/\partial \rho\right)_{\xi,S}\right]} \Bigg] \\
&\times \left[\left(\partial \rho/\partial A\right)_{\xi,S} + \rho_0^2 \left(\partial \xi/\partial \rho\right)_{A,S} \left(\partial \rho/\partial p\right)_{\xi,S}\right] \Bigg\}.
\end{aligned} \tag{2.84}$$

2.4.5 *Hydrodynamic equations with statistically independent variables*

The thermodynamic variables $x_1, x_2, \ldots, x_n$ are statistically independent if their cross-averages $\langle x_i x_j \rangle$ vanish for $i \neq j$, so that the probability of fluctuations is given by

$$P\left(\delta x_1, \delta x_2, \ldots, \delta x_n\right) = \exp\left[-\frac{A_1}{2}\left(\delta x_1\right)^2 - \frac{A_2}{2}\left(\delta x_2\right)^2 - \cdots - \frac{A_n}{2}\left(\delta x_n\right)^2\right] \tag{2.85}$$

with $\left\langle \left(\delta x_i\right)^2 \right\rangle = A_i^{-1}$.

For a one-component system, one can choose ρ, T or p, S as statistically independent variables, since $\langle \delta\rho\delta T\rangle = \langle \delta p \delta S\rangle = 0$. For a many-component system, there are different possibilities for choosing the set of the hydrodynamic variables $x_1, x_2, \ldots,$ which characterize the state of a system. Using well-known thermodynamic formulae [4], one can verify that there are different sets of statistically independent variables. We have chosen the mass fraction x, the pressure p, $div\,(v)$ and the variable $\phi = C_{p,x}T/T_0 - \alpha_T p/\rho_0$, where α_T is the isobaric thermal expansion and $C_{p,x}$ is the specific heat at constant pressure. Another possible set of statistically independent variables is the mass fraction x, the pressure p, $div\,(v)$ and the reduced entropy $S_1 = S - (\partial S/\partial x)_{p,T}\, x$, or the temperature T, the pressure p, $div\,(v)$ and the variable ϕ. Note that for a chemically reactive ternary mixture, it is convenient [57] to use the reduced entropy and the reduced concentration,

$$S_1 = S - (\partial S/\partial x)_{p,T}\, x_1; \quad x = x_1 - (\partial x_1/\partial x_2)_{p,T}\, x_2. \tag{2.86}$$

After performing a Fourier transform, the set of linearized hydrodynamic equations can be written symbolically in the form

$$\frac{d}{dt} N\,(k,t) = -M\,(k,t)\, N\,(k,t) \tag{2.87}$$

where $N\,(k,s)$ is a column vector with n hydrodynamic variables ($n = 4$ for the case discussed below), and the form of the $n * n$ matrix $\hat{M}(k,s)$ is defined by the hydrodynamic equations. Solutions of Eqs. (2.87) have the following form

$$N_i\,(k,s) = [\det M\,(k,s)]^{-1} \sum_j P_{ij}\,(k,s)\, N_j\,(k)\,, \tag{2.88}$$

where the P_{ij} are algebraic functions. From Eq. (2.88), one gets the correlation functions

$$\langle N_j\,(k,s)\, N_j\,(-k)\rangle = \frac{P_{ij}\,(k,s)}{\det\ M\,(k,s)} \left\langle |N_j\,(k)|^2 \right\rangle, \tag{2.89}$$

if all N_j are statistically independent.

An expression for the time-dependent correlation function is obtained by taking the inverse Laplace transform of Eq. (2.89),

$$\langle N_j\,(k,t)\, N_j\,(-k)\rangle = \frac{\left\langle |N_j\,(k)|^2 \right\rangle}{2\pi} \int_{-\infty}^{\infty} \frac{ds \exp\,(st)\, P_{ij}\,(k,s)}{\det M\,(k,s)}. \tag{2.90}$$

To perform the integration in (2.90), one has to find the roots of $\det M\,(k,s)$.

Neglecting the visco-reactive terms (setting $L_1 = 0$ in Eqs. (2.62)–(2.63)), one can write the complete set of linearized hydrodynamic equations in the variables p, x, and $div\,(v)$ and the new variable $\phi = C_{p,x}T/T_0 - \alpha_T p/\rho_0$. This choice of hydrodynamic variables has been used for non-reactive mixtures [58] , and then extended to reactive mixtures [59], [60]. These equations take the following form [60], [61]

$$\left(\frac{\partial \rho}{\partial x}\right)_{p,T} \frac{\partial x}{\partial t} + \left(\frac{\partial \rho}{\partial p}\right)_{x,T} \frac{\partial p}{\partial t} + \frac{T_0}{C_{p,x}} \left(\frac{\partial \rho}{\partial T}\right)_{x,p} \frac{\partial \phi}{\partial t} + \rho_0 div\,(v) = 0 \tag{2.91}$$

$$\frac{\partial div\,(v)}{\partial t} = -\nabla^2 p + \left(\frac{4}{3}\eta + \varsigma\right)\nabla^2 div\,(v)\,; \tag{2.92}$$

$$\frac{\partial x}{dt} = D\nabla^2 x + D\left(\frac{k_p}{p_0} + \frac{k_T\alpha_T}{\rho_0 C_{p,x}}\right)\nabla^2 p + D\frac{k_T}{C_{p,x}}\nabla^2\phi - L_0 A; \tag{2.93}$$

$$\begin{aligned}
&\rho_0 T_0\,(\partial\phi/\partial t) - \rho_0 k_T\,(\partial A/\partial x)_{p,T}\,(\partial x/\partial t)\\
&- T_0\left[\alpha_T - \rho_0\,(\partial S/\partial p)_{x,T}\right](\partial p/\partial t)\\
&= \lambda\left[(T_0/C_{p,x})\,\nabla^2\phi + (T\alpha_T/\rho_0 C_{p,x})\,\nabla^2 p\right].
\end{aligned} \tag{2.94}$$

The chemical term $L_0 A$ appears in Eq. (2.93). Since the affinity A vanishes at equilibrium, one can expand A in terms of x, p, and T,

$$A = \left(\frac{\partial A}{\partial x}\right)_{p,T} x + \left(\frac{\partial A}{\partial p}\right)_{x,T} p + \left(\frac{\partial A}{\partial T}\right)_{x,p} T. \tag{2.95}$$

Substituting $T = T_0\phi/C_{p,x} + T_0\alpha_T p/\rho_0 C_{p,x}$ into Eq. (2.95) yields

$$\begin{aligned}
L_0 A &= L_0\,(\partial A/\partial x)_{p,T}\left[x - (\partial x/\partial p)_{T,A}\,p\right.\\
&\quad\left. - (\partial x/\partial T)_{p,A}\,(T_0\phi/C_{p,x} + T_0\alpha_T p/\rho_0 C_{p,x})\right]\\
&= L_0\,(\partial A/\partial x)_{p,T}\left\{x - \left[(\partial x/\partial p)_{T,A} + (\partial x/\partial T)_{p,A}\,T_0\alpha_T/\rho_0 C_{p,x}\right]p\right.\\
&\quad\left. - (T_0/C_{p,x})\,(\partial x/\partial T)_{p,A}\,\phi\right\}.
\end{aligned} \tag{2.96}$$

Inserting (2.96) into (2.93) gives

$$\begin{aligned}
\partial x/dt = &\left[D\nabla^2 - L_0\,(\partial A/\partial x)_{p,T}\right]x\\
&+ \left\{D\,(k_p/p_0 + k_T\alpha_T/\rho_0 C_{p,x})\,\nabla^2\right.\\
&\left. - L_0\left[(\partial A/\partial p)_{x,T} + (\partial A/\partial T)_{x,p}\,T_0\alpha_T/\rho_0 C_{p,x}\right]\right\}p\\
&+ \left[\frac{Dk_T}{C_{p,x}}\nabla^2 + (\partial A/\partial x)_{p,T}\,(\partial x/\partial T)_{p,A}\,T_0 L_0/C_{p,x}\right]\phi.
\end{aligned} \tag{2.97}$$

Finally, inserting (2.97) into (2.94) yields

$$\partial\phi/\partial t = (\partial A/\partial x)_{p,T}\, k_T/T_0 \left[D\nabla^2 - L_0\,(\partial A/\partial x)_{p,T}\right] x$$
$$+ \left\{(\lambda/\rho_0 C_{p,x})\,\nabla^2 + k_T/T_0 \left[(Dk_T/C_{p,x})\,(\partial A/\partial x)_{p,T}\,\nabla^2\right.\right.$$
$$\left.\left. - (T_0 L_0/C_{p,x})\,(\partial A/\partial x)^2_{p,T}\,(\partial x/\partial T)_{p,A}\right]\right\}\phi$$
$$+ \left\{\left[\alpha_T/\rho_0 - (\partial S/\partial p)_{x,T}\right](\partial p/\partial t) + (\partial A/\partial x)_{p,T}\, k_T/T_0\right.$$
$$\times \left(\left[D\,(k_p/p_0 + k_T\alpha_T/\rho_0 C_{p,x})\,\nabla^2\right] - L_0\left[(\partial A/\partial p)_{x,T}\right.\right.$$
$$\left.\left.\left. + (\partial A/\partial T)_{x,p}\, T_0\alpha_T/\rho_0 C_{p,x}\right]\right) + (\lambda T_0\alpha_T/\rho_0 C_{p,x})\,\nabla^2\right\} p. \quad (2.98)$$

Equations (2.91), (2.92), (2.97) and (2.98) are the complete system of hydrodynamic equations.

For a binary mixture without a chemical reaction ($L_0 = 0$), the roots of the determinant of the matrix $M(k,t)$ and the appropriate correlation functions have been found to lowest order in k [61]. In many practical cases, there exist small parameters, such as

$$\frac{k\lambda}{v_s\rho_0 C_{p,x}}; \qquad \frac{k\,(4\eta/3+\zeta)}{3v_s\rho_0}; \qquad \frac{kD}{v_s}. \qquad (2.99)$$

To lowest order in these parameters, setting the determinant of the 4×4 matrix $\hat{M}(k,s)$ to zero yields two zeros and two sound modes with roots, $s_{1,2} = \pm i v_s k$. In the next higher-order approximation, there are two sound modes and two entangled heat conductivity-diffusion modes [61],

$$s_{1,2} = \pm i v_s k - \Gamma k^2, \qquad (2.100)$$

$$s_{3,4} = \frac{k^2}{2}\left[\lambda/\rho_0 C_{p,x} + D + (Dk_T/T_0 C_{p,x})^2\,(\partial A/\partial x)_{p,T}\right]$$
$$\pm \frac{k^2}{2}\left\{\left[\lambda/\rho_0 C_{p,x} + D + (Dk_T/T_0 C_{p,x})^2\,(\partial A/\partial x)_{p,T}\right]^2\right.$$
$$\left. - 4D\lambda/\rho_0 C_{p,x}\right\}^{1/2} \qquad (2.101)$$

$$\Gamma = \frac{1}{2\rho_0}\left\{\frac{4}{3}\eta + \zeta + (C_{p,A}/C_{v,x} - 1)\,\lambda/C_{p,x}\right.$$
$$\left. - \left[(\partial\rho/\partial x)_{p,T} + \frac{k_T}{C_{p,x}}(\partial\rho/\partial T)_{x,p}\,(\partial A/\partial x)_{p,T}\right]^2 \frac{D\,(\partial p/\partial\rho)_{x,T}}{\rho_0\,(\partial A/\partial x)_{p,T}}\right\}.$$
$$(2.102)$$

One can readily use the hydrodynamic equation including the chemical term ($L_0 \neq 0$) and perform the analogous calculations, but this procedure leads to very cumbersome expressions. Therefore, we consider the influence of a chemical reaction for the simplified case of a non-compressible liquid ($div\,(v) = 0$), neglecting fluctuations of pressure ($p = 0$) [60]. The complicated matrix $M\,(k,t)$ is then simplified, and the remaining hydrodynamic equations (2.97) and (2.98) have the following form

$$\frac{\partial x}{dt} + a_{11}x + a_{12}\phi = 0 \tag{2.103}$$

$$\frac{\partial \phi}{dt} + a_{21}x + a_{22}\phi = 0 \tag{2.104}$$

where

$$\begin{aligned} a_{11} &= Dk^2 + L_0\left(\frac{\partial A}{\partial x}\right)_{p,T}, \\ a_{12} &= Dk_Tk^2/C_{p,x} - (L_0T_0/C_{p,x})\left(\frac{\partial A}{\partial x}\right)_{p,T}\left(\frac{\partial x}{\partial T}\right)_{p,A}, \\ a_{21} &= \left(Dk_Tk^2/T_0\right)\left(\frac{\partial A}{\partial x}\right)_{p,T} + (k_TL_0/T_0)\left(\frac{\partial A}{\partial x}\right)^2_{p,T}, \\ a_{22} &= \lambda k^2/\rho_0 C_{p,x} + \left(k_T^2/T_0C_{p,x}\right)\left(\frac{\partial A}{\partial x}\right)_{p,T} Dk^2 \\ &\quad - (k_TL_0/C_{p,x})\left(\frac{\partial A}{\partial x}\right)^2_{p,T}\left(\frac{\partial x}{\partial T}\right)_{p,A}. \end{aligned} \tag{2.105}$$

It is convenient to express these elements by the characteristic times of diffusion τ_d, heat conductivity τ_h , and chemical reaction τ_c, defined by

$$\tau_d = \left(Dk^2\right)^{-1}; \; \tau_h = \left(\frac{\lambda}{\rho_0 C_{p,x}}k^2\right)^{-1}; \; \tau_c = \left[L_0\left(\frac{\partial A}{\partial x}\right)_{p,T}\right]^{-1}. \tag{2.106}$$

The matrix elements a_{ij} can then be rewritten,

$$\begin{aligned} a_{11} &= \tau_d^{-1} + \tau_c^{-1}; \quad a_{12} = \frac{k_T}{C_{p,x}}\tau_d^{-1} - \frac{T_0}{C_{p,x}}\left(\frac{\partial x}{\partial T}\right)_{p,A}\tau_c^{-1}, \\ a_{21} &= \frac{k_T}{T_0}\left(\frac{\partial A}{\partial x}\right)_{p,T}\tau_d^{-1} + \frac{k_T}{T_0}\left(\frac{\partial A}{\partial x}\right)_{p,T}\tau_c^{-1}, \\ a_{22} &= \tau_h^{-1} + \frac{k_T^2}{T_0C_{p,x}}\left(\frac{\partial A}{\partial x}\right)_{p,T}\tau_d^{-1} - \frac{k_T}{C_{p,x}}\left(\frac{\partial A}{\partial x}\right)_{p,T}\left(\frac{\partial x}{\partial T}\right)_{p,A}\tau_c^{-1}. \end{aligned} \tag{2.107}$$

The determinant Δ of the 2×2 matrix M_1 of Eqs. (2.103)–(2.104) has the following form

$$\Delta = (a_{11} - s)(a_{22} - s) - a_{12}a_{21}. \tag{2.108}$$

To solve Eqs. (2.103)–(2.104), one needs the roots s_1 and s_2 of Δ. The inverse Laplace transforms of the solutions of Eqs. (2.103) and (2.104), yields

$$\int_0^\infty dt \int_0^\infty ds \exp(st) \langle x(k,t)\, x(-k,0)\rangle = (a_{22}/s_1 s_2) \langle x(k,0)\, x(-k,0)\rangle,$$

$$\int_0^\infty dt \int_0^\infty ds \exp(st) \langle \phi(k,t)\, \phi(-k,0)\rangle = (a_{11}/s_1 s_2) \langle \phi(k,0)\, \phi(-k,0)\rangle,$$

$$\int_0^\infty dt \int_0^\infty ds \exp(st) \langle x(k,t)\, \phi(-k,0)\rangle = -(a_{12}/s_1 s_2) \langle \phi(k,0)\, \phi(-k,0)\rangle,$$

$$\int_0^\infty dt \int_0^\infty ds \exp(st) \langle \phi(k,t)\, x(-k,0)\rangle = -(a_{21}/s_1 s_2) \langle x(k,0)\, x(-k,0)\rangle. \tag{2.109}$$

The following calculations are analogous to those performed in Eq. (2.82), but instead of variables ξ and ρ, we use now the statistically independent variables x and ϕ. Using Eqs. (2.109), one obtains

$$\begin{aligned} L &= \left(L_0^2/\kappa_B T_0\right) \int_0^\infty dt \, \langle A(k,t)\, A(-k.0)\rangle \\ &= \frac{1}{s_1 s_2 \kappa_B T_0 \tau_c^2} \left\{ \left[\frac{a_{21} T_0}{C_{p,x}} \left(\frac{\partial x}{\partial T}\right)_{p,A} + a_{22} \right] \langle x(k,0)\, x(-k,0)\rangle \right. \\ &\quad \left. + \frac{T_0}{C_{p,x}} \left(\frac{\partial x}{\partial T}\right)_{p,A} \left[a_{12} + \frac{a_{11} T_0}{C_{p,x}} \left(\frac{\partial x}{\partial T}\right)_{p,A} \right] \langle \phi(k,0)\, \phi(-k,0)\rangle \right\}. \end{aligned} \tag{2.110}$$

Substituting into Eq. (2.110) the equilibrium correlators

$$\langle x(k,0)\, x(-k.0)\rangle = \frac{\kappa_B T_0}{(\partial A/\partial x)_{p,T}}; \; \langle \phi(k,0)\, \phi(-k.0)\rangle = \kappa_B C_{p,\xi} \tag{2.111}$$

one gets

$$L = \frac{1}{s_1 s_2 \tau_c^2} \left\{ \left[\frac{a_{21} T_0}{C_{p,x}} \left(\frac{\partial x}{\partial T} \right)_{p,A} + a_{22} \right] (\partial A/\partial x)_{p,T}^{-1} + \left(\frac{\partial x}{\partial T} \right)_{p,A} \left[a_{12} + \frac{a_{11} T_0}{C_{p,x}} \left(\frac{\partial x}{\partial T} \right)_{p,A} \right] \right\}. \tag{2.112}$$

According to Eq. (2.108), the product of the roots $s_1 s_2$ equals $a_{11}a_{22} - a_{12}a_{21}$. Substituting the latter into (2.112), and using Eq. (2.107), one finally obtains

$$L = \frac{L_0}{\tau_c} \left\{ \frac{1}{\tau_h} + \frac{k_T^2}{\tau_d T_0 C_{p,x}} \left(\frac{\partial A}{\partial x} \right)_{p,T} \left[1 + \frac{T_0}{k_T} \left(\frac{\partial x}{\partial T} \right)_{A,S} \right]^2 \right\} \times \left\{ \frac{1}{\tau_h} \left(\frac{1}{\tau_d} + \frac{1}{\tau_c} \right) + \frac{k_T}{\tau_d \tau_c C_{p,x}} \left[\left(\frac{\partial S}{\partial x} \right)_{p,T} + \frac{k_T}{T_0} \left(\frac{\partial A}{\partial x} \right)_{p,T} \right] \times \left[1 + \frac{T_0}{k_T} \left(\frac{\partial x}{\partial T} \right))_{A,S} \right] \right\}^{-1}. \tag{2.113}$$

The goal of our analysis was to find the mode coupling corrections to the Onsager coefficient, which define the change of the characteristic time of a chemical reaction $\tau_c = [L_0 (\partial A/\partial \xi)]^{-1}$. Such "mode coupling" is expressed by replacing the bare Onsager coefficient L_0 by the renormalized coefficient L, so that the effective characteristic time of a chemical reaction is $[L (\partial A/\partial \xi)]^{-1}$. In the general case, such a calculation involves a very cumbersome analysis. To avoid this, we consider some special cases.

When one neglects the thermal variables, viscous, diffusive and chemical modes appear in Eq. (2.84),

$$L = L_0^2 \left(\frac{\partial A}{\partial \xi} \right)_{p,S}^2 \left[\frac{L_1/\rho_0}{Dk^2 + L_0 (\partial A/\partial \xi)_{\rho,S}} + \frac{2\left(\frac{4}{3}\eta + \zeta\right) \rho_0 (\partial \xi/\partial \rho)_{A,S}}{\left(\frac{4}{3}\eta + \zeta\right)^2 k^2 + 4\rho_0 L_1 (\partial A/\partial \rho)_{\xi,S}} \right] \times \left[\left(\frac{\partial \rho}{\partial A} \right)_{\xi,S} + \rho_0^2 \left(\frac{\partial \xi}{\partial \rho} \right)_{A,S} \left(\frac{\partial \rho}{\partial p} \right)_{\xi,S} \right]. \tag{2.114}$$

Note that the diffusive mode appears in Eq. (2.114) because of the additional phenomenological visco-reactive coefficient L_1 introduced in Eqs. (2.59)–(2.60).

It is convenient to consider some limiting cases of Eq. (2.113):

Fast chemical reaction ($L_0\left(\frac{dA}{dx}\right)_{p,T} >> \frac{\lambda}{\rho_0 c_p}k^2 \approx Dk^2$):

$$L = \frac{L_0\left\{1 + \frac{\tau_h}{\tau_d}\frac{k_T^2}{T_0 C_{p,x}}\left(\frac{\partial A}{\partial x}\right)_{p,T}\left[\left(1+\frac{T_0}{k_T}\left(\frac{\partial x}{\partial T}\right)_{p,A}\right)^2\right]\right\}}{1+\frac{\tau_h}{\tau_d}\frac{k_T}{C_{p,x}}\left[\left(\frac{\partial S}{\partial x}\right)_{p,T}+\frac{k_T}{T_0}\left(\frac{\partial A}{\partial x}\right)_{p,T}\right]\left[1+\frac{T_0}{k_T}\left(\frac{\partial x}{\partial T}\right)_{p,A}\right]}. \tag{2.115}$$

If heat conductivity processes are faster than diffusive processes, ($L_0\left(\frac{dA}{d\xi}\right)_{p,T} >> \frac{\lambda}{\rho_0 c_{p,x}}k^2 >> Dk^2$), the presence of a chemical reaction does not change the Onsager coefficient, $L = L_0$. In the opposite limit, ($L_0\left(\frac{dA}{dx}\right)_{p,T} >> Dk^2 >> \frac{\lambda}{\rho_0 c_{p,x}}k^2$),

$$L = L_0\frac{(k_T/T_0)\left(\partial A/\partial x\right)_{p,T}\left(1+(T_0/k_T)\left(\partial x/\partial T\right)_{p,A}\right)}{\left(\partial S/\partial x\right)_{p,T}+(k_T/T_0)\left(\partial A/\partial x\right)_{p,T}}. \tag{2.116}$$

Fast heat conductivity ($\frac{\lambda}{\rho_0 c_{p,x}}k^2 >> L_0\left(\frac{dA}{dx}\right)_{p,T} \approx Dk^2$):

$$L = L_0\frac{L_0\left(dA/dx\right)_{p,T}}{L_0\left(dA/dx\right)_{p,T}+Dk^2}. \tag{2.117}$$

For fast diffusion processes ($\frac{\lambda}{\rho_0 c_{p,x}}k^2 >> Dk^2 >> L_0\left(dA/dx\right)_{p,T}$),

$$L = L_0\frac{L_0\left(dA/dx\right)_{p,T}}{Dk^2}, \tag{2.118}$$

whereas in the opposite limit, ($\frac{\lambda}{\rho_0 c_{p,x}}k^2 >> L_0\left(\frac{dA}{dx}\right)_{p,T} >> Dk^2$), $L = L_0$.

Fast diffusion ($Dk^2 >> \frac{\lambda}{\rho_0 c_{p,x}}k^2 \approx L_0\left(\frac{dA}{dx}\right)_{p,T}$):

$$\begin{aligned} L = {} & L_0^2\left(\frac{k_T^2}{T_0 C_{p,x}}\right)\left(\frac{\partial A}{\partial x}\right)_{p,T}^2\left[1+\frac{T_0}{k_T}\left(\frac{\partial x}{\partial T}\right)_{p,A}\right]^2 \\ & \times\left\{\frac{\lambda k^2}{\rho_0 c_{p,x}}+L_0\frac{k_T}{C_{p,x}}\left(\frac{\partial A}{\partial x}\right)_{p,T}\right. \\ & \left.\times\left[\left(\frac{\partial S}{\partial x}\right)_{p,T}+\frac{k_T}{T_0}\left(\frac{\partial A}{\partial x}\right)_{p,T}\right]\left[1+\frac{T_0}{k_T}\left(\frac{\partial x}{\partial T}\right)_{p,A}\right]\right\}^{-1}. \end{aligned} \tag{2.119}$$

If $Dk^2 >> \frac{\lambda}{\rho_0 c_{p,x}}k^2 >> L_0\left(\frac{dA}{dx}\right)_{p,T}$,

$$L = \frac{L_0^2\left(\frac{k_T^2}{T_0 C_{p,x}}\right)\left(\frac{\partial A}{\partial x}\right)_{p,T}^2\left[1+\frac{T_0}{k_T}\left(\frac{\partial x}{\partial T}\right)_{p,A}\right]^2}{\left(\lambda k^2/\rho_0 c_{p,x}\right)}, \tag{2.120}$$

whereas for $Dk^2 >> L_0 \left(\frac{dA}{dx}\right)_{p,T} >> \frac{\lambda}{\rho_0 c_p} k^2$,

$$L = L_0 \frac{\frac{k_T}{T_0}\left(\frac{\partial A}{\partial x}\right)_{p,T}\left[1 + \frac{T_0}{k_T}\left(\frac{\partial x}{\partial T}\right)_{p,A}\right]}{\left(\frac{\partial S}{\partial x}\right)_{p,T} + \frac{k_T}{T_0}\left(\frac{\partial A}{\partial x}\right)_{p,T}}. \tag{2.121}$$

Our hydrodynamic approach applies to the regime where k^{-1} is larger than the maximal characteristic size of a system. This is the correlation radius, which increases as the critical point is approached. Therefore, the first important consideration in the analysis of experimental data is the proximity to the critical point. Equations (2.114)–(2.121) also contain the thermodynamic derivatives and the kinetic coefficients, some of which change drastically near the critical points.

Even without analyzing specific reactive systems, one can see from Eqs. (2.114)–(2.121), that these equations contain all three types of behavior are described in [62]–[65] (slowing-down, speeding-up, unchanged).

2.5 Critical anomalies of chemical equilibria

In addition to singularities in the rate of reaction, peculiar behavior of equilibrium concentrations of reagents takes place near the critical point. The dissociation-recombination reaction of the form $B_2 \rightleftarrows 2B$ near the single liquid-gas critical point of the binary mixture of B_2 and B has been considered in [66], and improved in [67]. Common sense dictates that increasing (decreasing) the temperature should result in increasing (decreasing) the concentration of monomers in all dissociation-recombination reactions. Indeed, both energy (due to bond energy) and entropy (due to the number of particles) considerations seem to lead to the same conclusion. Quite unexpectedly, the opposite has been observed in qualitative experiments performed by Krichevskii and his collaborators [68] who studied the equilibrium reaction $N_2O_4 \rightleftarrows 2NO_2$ in a dilute solution of CO_2 near the critical point of the mixture. While another interpretation of these experiments exists [67], [69], it does not deny the existence of the anomalies in chemical equilibrium near the critical point. Consider varying the temperature along an equilibrium line of a reactive binary mixture $B_2 \rightleftarrows 2B$ at constant pressure. Since the affinity $A = \mu_{B_2} - 2\mu_B$ vanishes for any equilibrium state, along the equilibrium line one can write

$$dA = \left(\frac{\partial A}{\partial T}\right)_{\xi,p} dT + \left(\frac{\partial A}{\partial \xi}\right)_{T,p} d\xi = 0 \tag{2.122}$$

where $d\xi = dn_{B_2} - dn_B/2$ is the change in the extent of the reaction.

It follows from Eq. (2.122) that

$$\left(\frac{d\xi}{dT}\right)_{p,\ equil.line} = -\frac{(\partial H/\partial \xi)_{T,p}}{T\,(\partial A/\partial \xi)_{T,p}} \tag{2.123}$$

where the thermodynamic identity [3]

$$\left(\frac{\partial A}{\partial T}\right)_{\xi,p} = \left(\frac{\partial S}{\partial \xi}\right)_{T,p} = \frac{A + (\partial H/\partial \xi)_{T,p}}{T} \tag{2.124}$$

has been used, and H and S are the enthalpy and entropy, respectively. The derivative $(\partial H/\partial \xi)_{T,p}$ has no critical singularity. Therefore, the slope of concentration curves $\xi = \xi(T)$ is inversely proportional to the same derivative $(\partial A/\partial \xi)_{T,p}$ which determines the critical slowing-down. Hence, the derivative $(d\xi/dT)_{p,\ equil.line}$ exhibits a strong divergence at the critical point which is characterized [40] by critical indices γ or $(\delta - 1)/\delta$, depending on whether the critical point is approached at constant critical volume or pressure.

We used [70] similar arguments to explain the technologically important phenomenon of "supercritical extraction" — the substantial increase in the ability of near-critical fluids to dissolve solids. Fluids near critical points exist as a single phase having some of the advantageous properties of both a liquid and a gas. Indeed, they have the dissolving power comparable to a gas, transport properties intermediate between gas and liquid, and great sensitivity to small changes in pressure afforded by high compressibility.

Let s and f denote the solid phase and the fluid phase, respectively. The equality of the chemical potentials in the two phases is a necessary condition for the coexistence of these phases. Thus, for the solid,

$$\mu^s(T,p) - \mu^f(T,p,x) \equiv \Delta\mu = 0 \tag{2.125}$$

where we used the fact that the concentration x of the solid phase is equal to unity. Consider now isobaric changes in temperature along the equilibrium line,

$$d(\Delta\mu) = \left(\frac{\partial \mu^s}{\partial T}\right)_p dT - \left(\frac{\partial \mu^f}{\partial T}\right)_{p,x} dT - \left(\frac{\partial \mu^f}{\partial x}\right)_{T,p} dx = 0. \tag{2.126}$$

Equation (2.126) can be rewritten as

$$\left(\frac{\partial x}{\partial T}\right)_p = \frac{\overline{s} - s^s}{(\partial \mu^f/\partial x)_{T,p}} \tag{2.127}$$

where we introduce the molar entropy s of the solid $s^s = -(\partial \mu^s/\partial T)_p$ and the partial entropy of the solid in the fluid phase $\overline{s} = -\left(\partial \mu^f/\partial T\right)_{p,x}$.

Equations (2.126)–(2.127), which are analogous to (2.122)–(2.123), explain supercritical extraction. Indeed, near the critical points, the denominator in (2.127) approaches zero, leading to a strong divergence of the slope of the solubility curve.

The system described above contains a small amount of solid dissolved in a fluid. One may use the binary mixture as the solvent and the solid as the solute. We have suggested [71] the use for supercritical extraction in the neighborhood of a consolute point rather than at the liquid-gas critical point. The former is characterized by much lower pressures and temperatures, which has clear technological advantages.

2.6 Experiment

Experiments investigating the interplay between criticality and chemistry were performed long before the theorists showed any interest in this problem. The first publication of the oxidation reaction in the critical region goes back to 1946 [72]. Mention should be made of a series of qualitative experiments performed by the prominent Russian physical chemist Krichevskii and his collaborators [68]. In two experiments, they irradiated with light pure Cl_2 and I_2 in CO_2. In both cases, the irradiation results in the dissociation of the diatomic molecules into atoms. Under usual conditions, when the irradiation stops, rapid recombination restores chemical equilibrium. However, when the experiment is conducted under critical conditions (near the liquid-gas critical point of pure Cl_2 or of the binary CO_2-I_2 mixture), the recombination rate is dramatically reduced by one or two orders of magnitude. When the interpretation of these experiments led to some doubts [69], this problem came to the attention of the theorists, including myself. A few interesting experimental results were lately reported by S. Greer and collaborators. They measured the rate of recombination of chlorine atoms after thermal and chemical dissociation of chlorine near its liquid-vapor critical point [73]. They found that after a thermal perturbation, the recombination was very slow, whereas the recombination after the photolysis with ultraviolet light was very fast, contrary to the results obtained in [68]. Additional experiments have been performed [74] near the liquid-liquid critical point of a binary mixture with the aim of finding the change in the relative amounts of NO_2 and N_2O_4 during the dimerization of NO_2. Using the static dielectric constant as the measure of the extent of dimerization, they found 4% increase (and not decrease!) in dissociation near the critical point.

A review of early development in this field can be found in [75]. Special attention should be paid to the Snyder and Eckert article: "Chemical kinetics at a critical point" [76], where, in contrast to the well-known "slowing-down", they found "speeding-up" (increase by some 40%) of the reaction rate of the isoprene with maleic anhydride in the vicinity of the upper critical solution point of two liquid-liquid mixtures (hexane-nitrobenzene and hexane-nitroethane). This result was obtained more than twenty years before the theoretical idea of the "piston effect", and authors proposed an explanation related to the dilute solution of the reactants.

Extensive studies of the rate of chemical reactions in the vicinity of the critical point were carry out by J. K. Baird and collaborators. In their 1998 article [62], they described the measurement of the rate of five hydrolysis first-order reactions near the consolute point of three binary mixtures (tert-amyl chloride and tert-butyl chloride in isobutyric acid-water, 3-chloro-3-methylpentane and 4-methylbenzyl bromide in 2-butoxyethanol-water, tert-butyl bromide in triethylamine-water). All these reactions produce a strong electrolyte, and the progress of the reaction was followed by measuring the conductance in the vicinity of the lower and upper critical consolute points. All measurements showed a slowing-down effect within a few tenths of a degree on either side of the consolute temperature, in agreement with previous experiments. The measurement of 2-chloro-2-methylbutane near the critical point of the mixture isobutyric acid-water, has been repeated [63]. The more careful, later experiment (system was well-stirred in the two-phase region, and account was taken of the small shift $\Delta T_c = 0.2\ K$ of the critical temperature of the solvent due to the presence of the reactant) yields different results. Above T_C, the rate of reaction showed slowing-down, while below T_C, in the two-phase region, the rate of reaction increased upon approaching the critical point ("speeding-up" predicted by the "piston effect"). To check whether the speeding-up of the reaction rate is connected with the coexistence of two immiscible phases, the experiment was carried out on 2-bromo-2-methylpropane in triethylamine-water, analogous to that performed in [63]. In contrast to isobutyric acid-water used in [63], the latter mixture has a lower critical solution temperature. In spite of this difference, the results of the experiments were the same: slowing-down above the critical temperature and speeding-up below, which means that the "speeding-up" effect is associated with the critical point and not with the number of phases in the system.

Another improvement over the original experiments was that visual detection of critical opalescence to determine the critical temperature was replaced by the association of the critical temperature with the temperature of the onset of the sharp change in electrical conductivity when critical opalescence occurs [64]. The authors measured the rate of the Menschutkin reaction between benzyl bromide and triethylamine near the consolute point of triethylamine-water, which can be compared with the old measurements [76] of the Menshutkin reaction on ethyl iodide. For temperatures below T_C, they could clearly see the critical slowing-down of the reaction rate showing the weak singularity, while above T_C, a few measured points closest to T_C gave a hint of speeding-up. These results agree with those obtained earlier [76].

The next question investigated concerned the generality of the effect found. In particular, does the effect exist for other types of reactions, say, of second-order, and not of first-order kinetics? To answer this question, the experiment was performed [65] with the saponification of ethyl acetate by sodium hydroxide near the consolute point of the 2-butoxyethanol-water, which is known to exhibit [77] second-order kinetics. The experiment [65] shows slowing-down for temperatures below T_C. For $T > T_C$, the results of the experiment are not clear: one cannot a rule out speeding-up, but only future experiments will definitively answer this question.

An important comment has been emphasized in [62], [63] regarding the quantitative description of the slowing-down and speeding-up of the reaction rate. The rate of reaction is proportional to thermodynamic derivative of the Gibbs free energy difference ΔG separating products from reactance with respect to the extent of reaction ξ. As described in section 2.1, the singularity x of this quantity, $(\partial \Delta G / \partial \xi)_{a,b,\ldots} \sim |(T - T_C)/T_C|^x$, at the critical point depends on number of fixed "density type" variables a, b (entropy, volume, concentrations of the chemical components, extent of reaction). Namely, if one density is fixed, then one finds a "weak" singularity ($x = 0.11$ or $x = 0.33$, depending on the thermodynamical path approaching the critical point — fixed pressure or temperature or along the coexistence curve). For densities not held fixed, the singularity is "strong" ($x = 1.24$ or $x = 4.9$). Finally, for two or more fixed densities, $x = 0$ and there is no singularity of the reaction rate at the critical point. In turn, the number of fixed densities, i.e., those which do not participate in at least one chemical reaction, depends on the number of reactions occurring in the system. Estimating the number of reactions is not trivial, as seen from the following example [63]. One of the components of a solvent (isobutyric acid), which does not participate in the hydrolysis reaction, might nevertheless react with the alcohols present in the system, thereby changing the behavior of the reaction rate by changing the number of reactions and the number of densities held fixed.

In addition to conductance measurements, the same experimental group used another method to analyzing the progress of a chemical reaction [78], namely, the measurement of the rate of appearance of a phase that is not completely miscible with a liquid mixture, such as the gas phase. Three reactions have been studied: 1) acetone dicarboxylic acid in aniline catalyzed decomposition of acetone dicarboxylic acid in isobutyric acid-water, 2) benzene diazonium tetrafluoroborate in the decomposition of benzene diazonium tetrafluoroborate in 2-butoxyethanol-water, 3) ethyl diazoacetate in the acid catalyzed decomposition of ethyl diazoacetate in isobutyric acid-water. In all cases, the critical point did not effect the rates of the reactions.

Chapter 3

Effect of Chemistry on Critical Phenomena

3.1 Change of critical parameters due to a chemical reaction

Chemical reactions near the critical points are routinely studied near the consolute points of a binary mixture which plays the role of a solvent. The presence of a reactant, as well as impurities, pressure or electric field changes the critical parameters of a binary mixture. There is only limited knowledge of the relationship between the shifts of the critical parameters when the perturbation is applied. For example, it is known [3] that an impurity will cause a critical temperature shift equal to that of the critical composition only when the solubilities in the two components are roughly equal. However, it was found that two different impurities (water [79] and acetone [80]), which have quite different solubilities in the two components of the methanol-cyclohexane mixture, induce equal relative shifts in critical temperature and composition, $\Delta T_C/T_{C,0} = \Delta x_C/x_{C,0}$. The same relation has been observed [81] in a wide range of systems (Fig. 3.1). These data have been obtained by isobaric measurements upon the addition of impurities.

There are different experimental ways of finding the critical parameters T_C and p_C of a system. The simplest, but less precise method is based on visual detection of critical opalescence to determine the critical temperature. For ionic conductors, the critical temperature is identified by the sharp change in electrical conductivity [64]. Alternatively, one can use the dilatometric method of measuring the height of the liquid in the capillary side-arm as a function of temperature. A change in the slope of this curve occurs at the critical temperature [62]. Another method is the acoustic technique, using the fact that the velocity of sound reaches a minimum at the critical point [82]. Experiments show that the critical temperature T_C

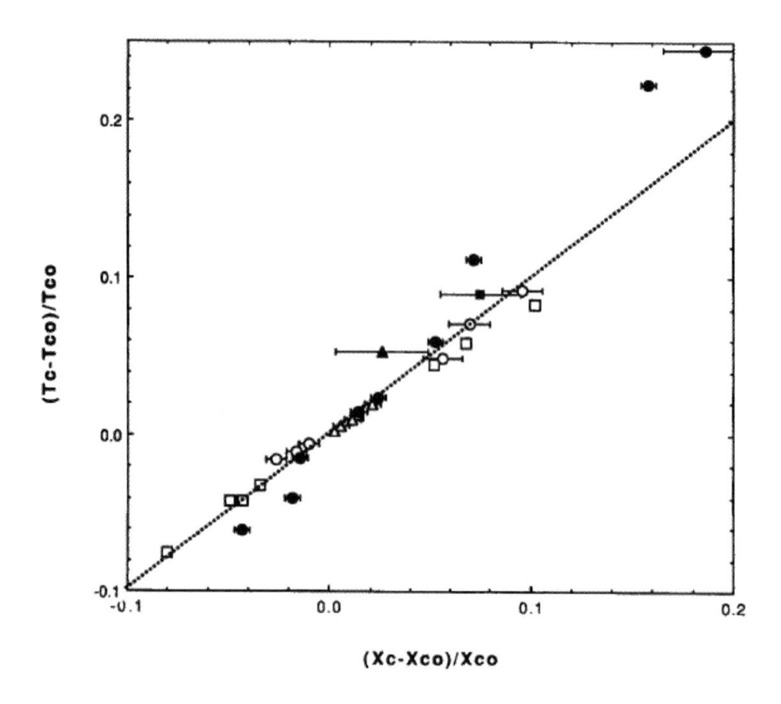

Fig. 3.1 The shift of critical points $T_{C,0}$ and $x_{C,0}$ of different systems to new values T_C and x_C. The different symbols refer to five different systems (for details, see [81]). Reproduced from Ref. [81] with permission, copyright (1989), American Institute of Physics.

is the linear function of the initial concentration x of a reactant,

$$T_C = T_C^0 + ax. \tag{3.1}$$

The sign of the constant a depends on whether the chemical reaction enhances or inhibits the mutual solubility of the original solvent pair. In the former case, $a < 0$ for UCST, and $a > 0$ for LCST, and vice versa for the latter case. For the equilibrium mixture of benzyl bromide in triethylamine-water $\left(T_C^0 = 291.24 \text{ K}\right)$, it was found [64] that $a = 36.9$ K/mole. There are many experimental results related to different many-component mixtures [82]. Measurements of the shift of the critical temperature in the presence of a chemical reaction have recently been performed for indium oxide dissolved in the near-critical isobutyric acid-water mixture [83].

3.2 Modification of the critical indices

In this section, we describe the changes in critical phenomena arising from the presence of chemical reactions. The singularities of the thermodynamic and kinetic quantities near the critical points are determined by the critical indices. The calculations of these indices assumes the constancy of some intensive variables, say, pressure p or chemical potentials μ, whereas in practice it is impossible, for example, to ensure the constancy of μ (one has to vary the concentrations during the course of the experiment). Similarly, p remains practically unchanged near the liquid-liquid critical point since the experiment is carried out in the presence of saturated vapor. On the other hand, it is quite difficult to ensure the constancy of p near the liquid-gas critical point where a fluid is highly compressible.

In order to compare theoretical calculations for an ideal one-component system with experimental data obtained for "real" systems, Fisher [84] developed the theory of renormalization of critical indices. The main idea of renormalization can be explained by the following simple argument. For a pure substance, mechanical stability is determined by $(\partial p/\partial v)_T < 0$. The condition for diffusion stability in a binary mixture has the form $(\partial \mu/\partial x)_{T,p} > 0$, which, by a simple thermodynamic transformation, can be rewritten as $(\partial p/\partial v)_{T,\mu} < 0$. Therefore, at constant chemical potential μ, the critical behavior of a binary mixture, which is defined by the stability conditions, will be the same as that of a pure substance. Multicomponent systems behave in an analogous manner, namely, the stability condition for a n-component mixture is determined by the condition $(\partial p/\partial v)_{T,\mu_1 \cdots \mu_{n-1}} < 0$, where $n-1$ chemical potentials are held constant. However, the critical parameters now depend on the variables μ_i, and in order to obtain experimentally observable quantities which correspond to constant concentration x, one has to pass from $T(\mu)$ to $T(x)$. As a result, when one goes from a pure substance to a binary mixture, the critical indices above and below a critical point are multiplied by $\pm(1-\alpha)^{-1}$ [84], where α is the critical index of the specific heat at constant volume. The minus sign refers to the specific heat at constant volume and the plus sign refers to all other critical indices. We have discussed [85] the renormalization of the critical indices associated with the presence of chemical reactions. As an example, let us again consider a binary fluid mixture $A_1 - A_2$ with the single reaction $\nu_1 A_1 + \nu_2 A_2 = 0$ (say, the isomerization reaction). This equation means that the law of mass action is satisfied, i.e., one gets for the extent of reaction $A = \nu_1 \mu_1 + \nu_2 \mu_2 = 0$, which, in turn, reduces by one the number of thermodynamic degrees of freedom in a binary mixture. Therefore, the reactive binary system has an isolated critical point, and the critical indices of this system are the same as those of a pure fluid,

$$\begin{aligned} \left(\frac{\partial v}{\partial p}\right)_{T,A=0} &\sim C_{p,A=0} \sim \left(\frac{T-T_C}{T_C}\right)^{-\gamma}; \\ \left(\frac{\partial v}{\partial p}\right)_{s,A=0} &\sim C_{v,A=0} \sim \left(\frac{T-T_C}{T_C}\right)^{-\alpha}, \text{ etc.} \end{aligned} \tag{3.2}$$

where $v = 1/\rho$ is the specific volume.

It is interesting to compare (3.2) with the case of a frozen chemical reaction (no catalyst added). The system considered is then a binary mixture with a liquid-gas critical line. The renormalized critical indices of such systems are well known [84]

$$\begin{aligned} C_{p,\xi} &\sim \left(\frac{\partial v}{\partial p}\right)_{T,\xi} \sim \left(\tfrac{T-T_C}{T_C}\right)^{-\alpha/(1-\alpha).}; \\ C_{v,\xi} &\sim \left(\frac{\partial v}{\partial p}\right)_{s,\xi} \sim \left(\tfrac{T-T_C}{T_C}\right)^{\alpha/(1-\alpha)}. \end{aligned} \tag{3.3}$$

The correspondence between (3.2) and (3.3) becomes obvious from the thermodynamic relation [3]

$$C_{p,A=0} = C_{p,\xi} - h^2 \left(\partial\xi/\partial A\right)_{T,p} \tag{3.4}$$

where h is the heat of reaction. In fact, the asymptotic behavior of the left-hand side of Eq. (3.4) on approaching a critical point is determined by the second term on the right-hand side rather than by the first term, which has a weaker singularity.

The occurrence of a chemical reaction in the system under consideration modifies the critical indices of the observable specific heats at constant volume and constant pressure (or inverse velocity of sound) compared to a system with a frozen chemical reaction. They are changed from $\alpha/(1-\alpha)$ and $-\alpha/(1-\alpha)$ to $-\alpha$ and $-\gamma$, respectively. For example, the specific heat at constant volume has a weak singularity at the critical point when a chemical reaction occurs, but a finite, cusped behavior in the absence of a chemical reaction.

The singularities become weaker for many-component mixtures. A common situation is that the solutes undergo various chemical transformations while the solvent does not. As an example, consider the critical system containing a reactive binary mixture dissolved in some solvent. Due to the existence of a neutral third component, the system has a line of critical points rather than an isolated critical point, as is the case for a reactive binary mixture.

Let us consider the frozen chemical reaction. The singularities of the thermodynamic quantities in a ternary mixture at constant chemical potential μ_0 of the solvent are similar to those of a binary mixture. The liquid-liquid critical points depend only slightly on the pressure, so that the parameter $R\rho_C\,(dT_C/dp)$ (ρ_C is the critical density) is very small. This parameter determines the region of renormalization [86]. Therefore, renormalization is absent, as in the vicinity of the λ line in helium, and

$$C_{p,\xi,\mu_0} \sim C_{v,\xi,\mu_0} \sim \phi\,(\mu_0)^{-\alpha}; \quad \phi\,(\mu_0) \equiv \frac{T - T_C\,(\mu_0)}{T_C\,(\mu_0)}. \tag{3.5}$$

However, measurements are taken at constant number of solvent particles rather than at $\mu_0 = const$. According to renormalization, near the liquid-liquid critical points, one obtains

$$C_{p,\xi,N_0} \sim C_{v,\xi,N_0} \sim \left(\frac{T - T_C}{T_C}\right)^{\alpha/(1-\alpha)}. \tag{3.6}$$

Unlike the liquid-liquid critical point, there are two renormalizations in the vicinity of liquid-gas critical point; the first occurs when passing from the binary to the ternary mixture, and the second renormalization takes place when passing from $\mu_0 = const$ to $N_0 = const$. In the region of renormalization of the ternary mixture, one gets

$$C_{p,\xi,\mu_0} \sim \phi^{-\alpha/(1-\alpha)}; \qquad C_{v,\xi,\mu_0} \sim \phi^{\alpha/(1-\alpha)} \tag{3.7}$$

and for $N_0 = const$,

$$C_{p,\xi,N_0} \sim \phi^{\alpha/(1-\alpha)}; \qquad C_{v,\xi,N_0} \sim const. \tag{3.8}$$

Comparing (3.6) and (3.8), one concludes that without a chemical reaction, cusp-like behavior exists for both specific heats near the liquid-liquid critical points, but only for the specific heat at constant pressure near the liquid-gas critical points.

When a chemical reaction takes place, the singularities near the critical points can be found from the thermodynamic relations, analogous to (3.4),

$$\begin{aligned} C_{p,N_0,A=0} &= C_{p,N_0,\xi} - T\,(\partial\xi/\partial A)_{T,p,N_0}\,(\partial A/\partial T)^2_{p,N_0,\xi} \\ C_{v,N_0,A=0} &= C_{v,N_0,\xi} - T\,(\partial\xi/\partial A)_{v,T,N_0}\,(\partial A/\partial T)^2_{v,N_0,\xi}\,. \end{aligned} \tag{3.9}$$

For both types of critical points, the singularities of thermodynamic quantities in a system undergoing a chemical reaction ($A = 0$) are determined by the second term on the right-hand sides of Eqs. (3.9). The factor

$\partial A/\partial T$ in the latter terms remains finite at the critical point, whereas the second factor has the following asymptotic behavior,

$$(\partial\xi/\partial A)_{T,p,\mu_0} \sim \phi^{-\gamma}; \qquad (\partial\xi/\partial A)_{v,T,\mu_0} \sim \phi^{-\alpha} \tag{3.10}$$

and, after renormalization,

$$\begin{aligned}(\partial\xi/\partial A)_{T,p,N_0} &\sim \left(\frac{T-T_C}{T_C}\right)^{-\alpha/(1-\alpha)}; \\ (\partial\xi/\partial A)_{v,T,N_0} &\sim \left(\frac{T-T_C}{T_C}\right)^{\alpha/(1-\alpha)}.\end{aligned} \tag{3.11}$$

Therefore, for a system undergoing a chemical reaction, one obtains from (3.9) and (3.11),

$$\begin{aligned}C_{p,N_0,A=0} &\sim \left(\frac{\partial v}{\partial p}\right)_{T,N_0,A=0} \sim \phi^{-\alpha/(1-\alpha)}; \\ C_{v,N_0,A=0} &\sim \left(\frac{\partial v}{\partial p}\right)_{v,N_0,A=0} \sim \phi^{\alpha/(1-\alpha)}.\end{aligned} \tag{3.12}$$

Comparing (3.6) and (3.8) with (3.12) yields that the existence of a chemical reaction leads to a magnification of the singularities of the specific heats at constant pressure for both types of critical points, from cusp-like behavior $\alpha/(1-\alpha)$ to a weak singularity $-\alpha/(1-\alpha)$. On the other hand, the specific heat at constant volume changes its asymptotic behavior from constant to cusp-like only near the liquid-gas critical point. In principle, one can detect such a magnification experimentally. Experiments are slightly easier to perform near the liquid-liquid critical points, because these points usually occur at atmospheric pressure and room temperature.

Consider now the singularities of the dielectric constant ε and electrical conductivity σ at the critical point of reactive systems. These quantities can be obtained by choosing poorly conducting liquid mixtures for dielectric measurements and strongly conducting mixtures for conductivity measurements. One can also measure ε and σ simultaneously placing the fluid between two parallel plates or coaxial cylindrical electrodes with an alternative current bridge. The fluid then acts in the bridge circuit as a frequency-dependent impedance with balanced capacity and resistive components [87]. Near the critical points, large fluctuations affect all thermodynamic and kinetic properties including those which are characterized by σ and ε. There are hundreds of experimental papers, of which we will describe only a few. Of special importance are chemical reactions in ecological clean near-critical water. At high temperature and pressure, water becomes self-ionized with a high conductivity, resembling molten salts [88]. The special case is the dissociation of a weak acid in water near an acid-water liquid-liquid critical point,

$$HA+H_2O \rightleftarrows A^-+H_3O^+. \tag{3.13}$$

Measurements of the conductivity as a function of temperature have been performed near the liquid-liquid critical points for the systems isobutyric acid-water [89], [90] and phenol-water [90]. Different explanations have been proposed for the anomaly in the conductivity near the critical point [91]: 1) The anomaly is due to viscous drag of the acid anion [89] and, hence, according to the Stokes law, it is proportional to the inverse of the viscosity η, $\sigma \sim \eta^{-1} = \tau^{\nu z_\eta} = \tau^{0.031}$, where $\tau = (T - T_C)/T_C$, and $\nu = 0.63$, $z_\eta = 0.05$ are the critical indices for the correlation length and viscosity, respectively; 2) The anomaly of σ is related to the anomaly in the proton-transfer rate [92], which, in turn [93], has the temperature dependence of the nearest-neighbor correlation function, $\sigma \sim \tau^{1-\alpha} = \tau^{0.89}$, where $\alpha = 0.11$ is the critical index for the heat capacity at constant pressure and composition; 3) A percolation model for conductivity leads [90] to an anomaly of the form $\sigma \sim \tau^{2\beta} = \tau^{0.65}$, where $\beta = 0.325$ is the critical index describing the coexistence curve; 4) The anomaly of σ is related to that of the extent of the acid dissociation [94], which leads [95] to $\sigma \sim \tau^{1-\alpha} = \tau^{0.89}$.

A detailed analysis of the experimental data obtained in [89], [90] has been carried out [96], taking account of the background contribution due to the normal temperature dependence of the dissociation constant and due to the confluent critical singularity. The leading critical anomaly of the conductance was found to be characterized by a critical index $1 - \alpha$, consistent with its connection with an anomaly in the extent of the dissociation reaction and also for an anomaly in the proton-transfer rate.

The static dielectric constant for the dimerization reaction $2NO_2 \rightleftarrows N_2O_4$ has been measured [74] in a solvent mixture of perfluoromethylcyclohexane-carbon tetrachloride near the liquid-liquid critical point of the solvent. The appropriate choice of reaction (no dipole moment for the N_2O_4 molecule, and a small moment for NO_2) and of the solvent (the components of low polarity make the "background" dipole moment very small) assure the high precision of measurements. The dielectric constant was found to have a $1-\alpha$ anomaly near the critical point, the same as the conductivity. For a mixture of polystyrene and diethyl malonate near the liquid-liquid critical point in the absence of chemical reactions, the dielectric constant was measured [97] as a function of frequency in the range 20 KHz to 1 MHz. No anomaly was found in the dielectric relaxation time.

3.3 Singularity in the degree of dissociation near a critical point

At the end of the previous section, we considered the anomalies in electroconductivity near the critical point, which are caused by the scattering of charged ions by strongly enhanced concentration fluctuations. However, apart from the anomalies of scattering processes, an additional factor has to be taken into account, namely, the density of ions may have a singularity near the critical point due to the chemical reaction [94]. Assume that the dissociation reaction

$$\mathrm{CD} \rightleftarrows \mathrm{C}^{+} + \mathrm{D}^{-} \tag{3.14}$$

occurs between the solutes CD, C^{+} and D^{-} near the critical point of the solvent E. The condition for chemical equilibrium has the following form,

$$A \equiv \mu_{CD} - \mu_{C^+} - \mu_{D^-} = 0. \tag{3.15}$$

The constraint of electroneutrality

$$N_{C^+} = N_{D^-}, \tag{3.16}$$

permits one to write the differential of the Gibbs free energy in the form

$$dG = -SdT + Vdp + \mu_E dN_E + \mu_{CD} dN_{CD} - (\mu_{C^+} + \mu_{D^-})\, dN_{C^+}. \tag{3.17}$$

It is convenient [95] to express the thermodynamics in terms of the total number N_1 of moles of CD, both dissociated and undissociated, and of total number N_2 of moles of CD and E,

$$N_1 = N_{CD} + N_{C^+}; \qquad N_2 = N_1 + N_E. \tag{3.18}$$

It is also convenient to define intensive variables,

$$\xi = \frac{N_{C^+}}{N_2}; \ \chi = \frac{N_1}{N_2}; \ g = \frac{G}{N_2}; \ v = \frac{V}{N_2}; \ s = \frac{S}{N_2}. \tag{3.19}$$

Using Eqs. (3.15)–(3.19), one may rewrite Eq. (3.17),

$$dg = -sdT + vdp - Ad\xi + (\mu_{CD} - \mu_E)\, d\chi. \tag{3.20}$$

Note that for a closed system, the variable χ remains constant, independent of the extent of reaction ξ. Usually the temperature changes during the experiment while the pressure remains constant. The extent of reaction, i.e., the number of charges, is defined by the derivative $(\partial\xi/\partial T)_{p,\chi,A=0}$. This derivative is taken under the condition that two extensive ("field") variables, p and A, and one intensive ("density") variable χ are kept fixed.

According to the general theory [38], this derivative has only a "weak" singularity at the critical point, defined by the critical index $\alpha = 0.12$,

$$\left(\frac{\partial \xi}{\partial T}\right)_{p,\chi,A=0} \approx \left(\frac{T - T_C}{T_C}\right)^{-\alpha}. \qquad (3.21)$$

Thus far, we have considered a one-phase system. Let us now consider a ternary mixture (components 1, 2 and 3) existing in two phases, which we label α and β. The chemical reaction allowed is of the form $1 \rightleftarrows 2 + 3$. For example, this model describes a one-component system (say HI) near its liquid-gas critical point ($T_C = 150$ K for HI) dissociated into two species ($2HI \rightleftarrows H_2 + I_2$). The question is: How will the concentration of the reagents vary near the critical point of a three-component system? For a small equilibrium constant of reaction, we may neglect the small shift from the critical point of a one-component system.

Denoting by μ_i^α and μ_i^β the chemical potentials of the i-th component in each phase, one gets for the equilibrium state,

$$\mu_1^\alpha = \mu_1^\beta; \quad \mu_2^\alpha = \mu_2^\beta; \quad \mu_3^\alpha = \mu_3^\beta. \qquad (3.22)$$

The condition of chemical equilibrium, similar to Eq. (3.15), has the form

$$\mu_1^\alpha - \mu_2^\alpha - \mu_3^\alpha = 0. \qquad (3.23)$$

A similar condition in the second phase follows from (3.22), but does not produce any additional restriction.

We are interested in equilibrium properties. Therefore, the thermodynamic path we choose is a displacement along the equilibrium line, keeping the pressure constant,

$$\begin{aligned} d\left(\mu_1^\alpha - \mu_1^\beta\right) &= d\left(\mu_2^\alpha - \mu_2^\beta\right) = d\left(\mu_3^\alpha - \mu_3^\beta\right) \\ &= d\left(\mu_1^\alpha - \mu_2^\alpha - \mu_3^\alpha\right) = 0. \end{aligned} \qquad (3.24)$$

Choosing as independent variables the mole fractions of the first and the second component in each phase, x_1^α, x_2^α, x_1^β, x_2^β, and the temperature T, one can rewrite Eqs. (3.24) in the following forms

$$\begin{aligned} \mu_{11}^\alpha dx_1^\alpha + \mu_{12}^\alpha dx_2^\alpha - \mu_{11}^\beta dx_1^\beta - \mu_{12}^\beta dx_2^\beta &= \left(\mu_{1T}^\beta - \mu_{1T}^\alpha\right) dT, \\ \mu_{21}^\alpha dx_1^\alpha + \mu_{22}^\alpha dx_2^\alpha - \mu_{21}^\beta dx_1^\beta - \mu_{22}^\beta dx_2^\beta &= \left(\mu_{2T}^\beta - \mu_{2T}^\alpha\right) dT, \\ \mu_{31\partial x}^\alpha dx_1^\alpha + \mu_{32}^\alpha dx_2^\alpha - \mu_{31}^\beta dx_1^\beta - \mu_{32}^\beta dx_2^\beta &= \left(\mu_{3T}^\beta - \mu_{3T}^\alpha\right) dT, \\ \left(\mu_{11}^\alpha - \mu_{21}^\alpha - \mu_{31}^\alpha\right) dx_1^\alpha + \left(\mu_{12}^\alpha - \mu_{22}^\alpha - \mu_{32}^\alpha\right) dx_2^\alpha &= -\left(\mu_{1T}^\alpha - \mu_{2T}^\alpha - \mu_{3T}^\alpha\right) dT \end{aligned} \qquad (3.25)$$

where

$$\left(\frac{\partial \mu_i^{\alpha,\beta}}{\partial x_j^{\alpha,\beta}}\right)_{p,T,x_i\neq x_j} \equiv \mu_{i,j}^{\alpha,\beta}; \quad \left(\frac{\partial \mu_i^{\alpha,\beta}}{\partial T}\right)_{p,x_j} \equiv \mu_{i,T}^{\alpha,\beta}. \tag{3.26}$$

Solutions of Eqs. (3.25) for $dx_i^{\alpha,\beta}/dT$ are

$$\left(\frac{dx_i^{\alpha,\beta}}{dT}\right)_{p,equil.} = \frac{\Delta_i^{\alpha,\beta}}{\Delta} \tag{3.27}$$

where the subscript equil refers to the equilibrium line, Δ is the 4×4 determinant of Eqs. (3.25), and $\Delta_i^{\alpha,\beta}$ is the determinant with one of the rows replaced by the row of coefficients from the right-hand side of Eqs. (3.25). Simple analysis shows [94] that Δ has no singularities at the critical point, whereas $\Delta_i^{\alpha,\beta}$ is proportional to the factor $\mu_{11}^{\alpha,\beta}\,\mu_{22}^{\alpha,\beta} - \left(\mu_2^{\alpha,\beta}\right)^2$ which can be rewritten as $\left(\partial \mu_{12}^{\alpha,\beta}/\partial x\right)_{p,T,\mu_1(p,T,x_1)}$. The line of critical points is thus determined by the vanishing of the next derivative of $\mu_2^{\alpha,\beta}$ with respect to x [3], which, having one "density" variable fixed, has a weak singularity (3.21) at the critical point. Then, the x-T curves have an infinite slope at the critical points.

The coincidence of the final results for the two cases results from the electroneutrality of the four-component mixture which makes it isomorphic to the three-component mixture. If the four-component mixture is comprised of two phases, each phase can be described by two mole fractions, say x_1 and x_2, while $x_3 = x_4 = (1 - x_1 - x_2)/2$. Therefore, the equilibrium equations will have the same form, Eq. (3.25), resulting in Eq. (3.21).

3.4 Isotope exchange reaction in near-critical systems

The results of experiments performed on isobutyric acid (CO_2H) - deuterium oxide (D_2O) binary mixture are an example of the shift of critical parameters and the modification of critical indices due to a chemical reaction. Peculiar to this system is the capability of the proton of the carboxyl group of isobutyric acid to exchange with the deuteron of deuterium oxide (isotope exchange reaction). This reaction produces more than two species, and the concentrations of CO_2H and D_2O are not the same because of the isotope exchange reaction. The experiments include the visual observation of the phase separation and light scattering [98]. The results have been compared with those for isobutyric acid-water (CO_2H/H_2O) and deuterated isobutyric acid-deuterium oxide (CO_2D/D_2O) systems. Appreciable amounts of additional species due to isotope exchange distort the coexistence curve, shifting the critical solution concentration x_C away from the concentration x_0 where the maximum phase separation temperature $T_{\max}$ occurs. This makes the critical indices γ and ν of the correlation length ($\xi = \xi_0 \left(\frac{T-T_C}{T_C}\right)^{-\gamma}$) and isothermal compressibility ($k = k_0 \left(\frac{T-T_C}{T_C}\right)^{-\nu}$) in the one-phase region ($T > T_C$) different from those of the coexisting two-phase region ($T < T_C$). It was found that x_C (mass fraction of isobutyric acid) equals 0.358 at $T_C = 44.90°C$, which differs from $x_0 = 0.310$ and $T_{\max} = 45.11°C$. In the one-phase region, $\gamma = 1.25$, $\nu = 0.63$ and $\xi_0 = 3.13$ Å in agreement with the renormalization group results, whereas in the coexistence two-phase region, the critical indices appear to be renormalized with $\gamma = 1.39, \nu = 0.76$ and $\xi_0 = 0.6$ Å. As result of exchange equilibria, CO_2H/D_2 O is effectively a three-component system in the coexistence two-phase region [98], and the renormalized critical indices are in agreement with those near the plait point of a ternary liquid mixture ethanol-water-chloroform [99]. An alternative explanation of these experiments is given in [100], [101].

3.5 Singularities of transport coefficients in reactive systems

In order to determine the influence of a chemical reaction on the transport coefficients, one can use the same mode-mode coupling analysis [102] used in the previous chapter for finding the influence of criticality on the rate of a chemical reaction. That is, one calculates the correlation functions which, in turn, define the change of the transport coefficients. A different way of attacking this problem is by dynamic renormalization group methods [103], which we will briefly describe in the next section.

3.5.1 *Mode-coupling analysis*

Just as for the reaction rates, the transport coefficients (shear viscosity η and diffusion coefficient D) can be written as time integrals of the correlations of appropriate fluxes, which, in turn, can be expanded in powers of the hydrodynamic variables. In calculating the critical corrections $\Delta\eta$ and ΔD, one neglects the small critical indices, thus assuming the Ornstein-Zernike form for the susceptibility. The most significant contribution to $\Delta\eta$ is the integral over wavenumber k of the form [36]

$$\Delta\eta \sim \int_{k_1}^{k_2} \frac{dk}{k+L}. \tag{3.28}$$

The upper limit k_2 is related to an unimportant cutoff, and the lower limit k_1, the inverse correlation length ξ^{-1}, vanishes at the critical point. When there is no reaction (Onsager coefficient $L = 0$), this integral diverges logarithmically at the critical point. A more careful calculation of $\Delta\eta$ leads to a weak power divergence rather than a logarithmic divergence. However, when $L \neq 0$, the integral (3.28) does not diverge. It follows that the viscosity in reactive systems remains finite at the critical point. Note that if L is extremely small (extremely slow chemical reaction), then η grows upon approaching the critical point, but its growth is terminated when the correlation length becomes of order L^{-1}.

The calculation of the mode-coupling corrections to the wavenumber (frequency)-dependent diffusion coefficient in reactive binary mixtures [102], [36] yields

$$\Delta D = \frac{\kappa_B T}{6\pi\rho\xi X^2\eta}\left[K(X) - \frac{\beta X^3}{Y^3}K(Y)\right] \tag{3.29}$$

where ξ is the correlation length, $X = k\xi$, $Y = X\sqrt{\delta/(1+X+\delta)}$, $\delta = L/\eta$, $\beta = \delta/(1+\delta)$. The function $K(X)$ is given by

$$K(X) = \frac{3}{4}\left[1 + X^2 + \left(X^3 - X^{-1}\right)\tan^{-1} X\right]. \tag{3.30}$$

When $L = 0$, one regains the well-known result for non-reactive mixtures [102]. The influence of the chemical reaction on the diffusion coefficient can easily be seen in the two limiting cases of low frequencies $k\xi < 1$ (the "hydrodynamic" region) and high frequencies $k\xi > 1$ (the "critical" region).

a) For $k\xi < 1$, $K(X) \sim X^2$, and

$$\Delta D = \frac{\kappa_B T}{6\pi\rho\xi\eta}\left(1 - \beta^{1/2}\right). \tag{3.31}$$

The usual result [102] $\eta\Delta D \sim \xi^{-1}$ is obtained, but with a renormalized coefficient. Because $D = \alpha\left(\partial A/\partial\xi\right)_{T,p}$, and since η is not divergent in reactive systems, α diverges like ξ.

b) For $k\xi > 1$, $K(X) \sim (3\pi/8)X^2$, and

$$\Delta D = \frac{\kappa_B T\, k}{16\rho(\eta + L)}. \tag{3.32}$$

This result is similar to that obtained for a nonreactive binary mixture, except for the renormalization of the coefficient of k. Therefore, the diffusion coefficient in reactive systems still vanishes like ξ^{-1} but with a modified coefficient of proportionality due to the chemical reaction.

3.5.2 *Renormalization group methods*

The hydrodynamic equations (2.61)–(2.64) are written for local space-averaged variables. The renormalization group technique allows one to average over increasingly large regions, thereby reducing the number of effective degrees of freedom of the system. Due to the divergence of the correlation length, such a procedure is of special importance near the critical points, where the increasing number of degrees of freedom becomes important. Efficient computer programs have been developed for studying static critical phenomena [104], and they have been extended to cover dynamic critical phenomena as well [103].

The starting point is the identification of slow modes, whose relaxation times approach zero at small wavenumbers, and also (if the order parameter is not conserved) the order-parameter mode. Therefore, one has to assign each system to the appropriate model $A - H$ [103]. Such an identification also depends on the regimes of wavenumber which define the relative importance of the different modes.

The distinctive feature of reactive systems is the existence of a homogeneous chemical mode, which converts the concentration into a non-conserved parameter. Milner and Martin [40] performed a renormalization group analysis for a reactive binary mixture, improving the result of the linear analysis of the critical slowing-down described in Sec. 2.1.

They found [40] that the critical slowing-down of a chemical reaction occurs for $k_H < k < k_C$, where k_C and k_H are the inverse length scales for diffusion and heat conductivity. The slowing-down is governed by the strong critical index $\gamma + \alpha + \eta\nu \simeq 1.37$ rather than by $\gamma \approx 1.26$ obtained in the linear theory.

Chapter 4

Phase Separation in Reactive Systems

4.1 Multiple solutions of the law of mass action

Thus far, we have considered single-phase reactive systems near their critical point. Another interesting problem is phase separation in reactive systems as compared to non-reactive systems. Will the presence of a chemical reaction, induced, say, by a very small amount of catalyst, stimulate or restrict the process of phase separation? In this chapter we will consider these problems, but we first have to consider the possibility of multiple solutions of the law of mass action. Two (or more) coexisting phases which have the same temperature and pressure are different in concentrations of different components and, therefore, in conducting chemical reactions. This means that the law of mass action has more than one solution for the concentration at given temperature and pressure.

Let us clarify the possibility of multiple solutions of the law of mass action. The chemical potential of the i-th component can be written in the following form

$$\mu_i = \mu_i^0 \left(p, T\right) + \kappa_B T \ln \left(\gamma_i, x_i\right) \tag{4.1}$$

where the activity γ_i determines the deviation from the ideal system, for which $\gamma_i = 1$. Using (4.1), one can rewrite the law of mass action as

$$x_1^{\nu_1}, x_2^{\nu_2}, \ldots, x_n^{\nu_n} = K_{id.} \left(p, T\right) \; \gamma_1^{-\nu_1}, \gamma_2^{-\nu_2}, \ldots, \gamma_n^{-\nu_n} \equiv K. \tag{4.2}$$

The chemical equilibrium constant for the ideal system K_{id} is determined by the functions μ_i^0 in (4.1), i.e., by the properties of individual non-reactive components, whereas for non-ideal systems, K also depends on the interactions among the components. For ideal systems, all $\gamma_i = 1$, and Eq. (4.2) has a single set of solutions $x_1, x_2, \ldots, x_n$. This was proved many years ago [105] for isothermal-isochoric systems. Another proof was given for isobaric systems under isothermal [106] and adiabatic [107] conditions. A complete analysis has recently been carried out [108].

It worth mentioning that phase separation is possible in ideal, although slightly artificial systems, such as that for which n units (atoms, molecules) of type A can reversibly form an aggregate (molecule, oligomer) of type A_n. Then, for large n (in fact, $n \to \infty$), the law of mass action has more than one solution, indicating the possibility of phase separation. It was subsequently noticed [109] that the latter effect is a special case of the phenomenon of the enthalpy-entropy compensation. This implies that for chemical reactions exhibiting a linear relationship between enthalpy and entropy, the magnitude of change in the Gibbs free energy is less than one might expect. Mathematically, in the Gibbs free energy equation ($\Delta G = \Delta H - T\Delta S$), the change in enthalpy (ΔH) and the change in entropy (ΔS) have opposite signs. Therefore, ΔG may change very little even if both enthalpy and entropy increase. However, the existence of the enthalpy-entropy compensation effect has been doubted by some researchers, as one can see from the article entitled "Enthalpy–entropy compensation: a phantom phenomenon" [110].

In general, a system has to be considerably non-ideal for the existence of multiple solutions of the law of mass action (4.2), i.e., the interaction energy must be of order of the characteristic energy of an individual component. One possibility is for the interaction energy between components to be high. Consider a gas consisting of charged particles (plasma, electrolytes, molten salts, metal-ammonium solutions, solid state plasma). Another possibility occurs when the characteristic energies of single particles are small, as in the case of isomers. Both the isomerization [111] and the dissociation [112] reactions have been analyzed in detail.

Consider the ionization equilibrium of the chemical reaction of the form $A \rightleftarrows i + e$ (dissociation-recombination of neutral particles into positive and negative charges). Neglecting the complications associated with the infinite number of bound states, assume that $K_{id} \sim \exp(I_0/\kappa_B T)$, where I_0 is the ionization potential of a neutral particle. Assuming Debye screening of the electrostatic interactions, one readily finds [112], [113] that

$$K = K_{id} \exp\left[-\Phi(x,T)/\kappa_B T\right] \approx \exp\left\{\left[I_0 - \Phi(x,T)\right]/\kappa_B T\right\}$$
$$\Phi(x,T) \approx (8\pi x)^{1/2}\left(e^2/\kappa_B T\right)^{3/2} + Bx + \cdots \tag{4.3}$$

where x is the concentration of charges, and the function B, which determines the pair correlation between charges, has been tabulated [106].

At certain T and p, Eq. (4.3) has several solutions for x. This has a simple physical meaning. The function $\Phi(x,T)$ in (4.3) diminishes the ionization potential I_0 as a result of screening. Therefore, the phase with the larger degree of ionization has higher energy and higher entropy. Hence, these two phases can have equal chemical potentials, and therefore, can co-exist. The chemical reaction proceeds in different ways in the two coexisting phases. Therefore, the appearance of multiple solutions of the law of mass action is a necessary condition for phase separation in reactive systems, where a chemical reaction proceeds in all phases.

4.2 Phase equilibrium in reactive binary mixtures quenched into a metastable state

4.2.1 *Thermodynamic analysis of reactive binary mixtures*

Let us assume that two components A_1 and A_2 participate in a chemical reaction of the form $\nu_1 A_1 \longleftrightarrow \nu_2 A_2$. The law of mass action for this reaction is

$$A = \nu_1 \mu_1 + \nu_2 \mu_2 = 0 \tag{4.4}$$

where A is the affinity of reaction. If the system separates into two phases, their temperatures, pressures and chemical potentials are equal

$$\mu_i(p,T,x') = \mu_i(p,T,x''); \qquad i = 1,2 \tag{4.5}$$

where x' and x'' denote the concentrations of one of the components in the two phases.

The kinetics of phase separation from O_2 into two phase, described by points O_3 and O_4 in Fig. 4.1, proceeds in two clearly distinguishable stages. During the first stage, the system is located at point O_2, "waiting" for the appearance of a significant number of critical nuclei due to fluctuations (the duration of this stage is usually called "the lifetime of the metastable state"). At the end of this stage, the system starts to separate into two phases, and after some "completion" time, it reaches the two-phase state described by the points O_3 and O_4.

The existence of a chemical reaction(s) increases the stability of a metastable state through the appearance of additional constraint(s). Moreover, thermodynamic considerations give only necessary but not sufficient conditions for phase separation in reactive mixtures. Even when thermodynamics allows phase separation, it might be impossible from the kinetic aspect. The latter can be achieved in two different ways. First, if the forward and backward reaction rates are different, with a preferred component

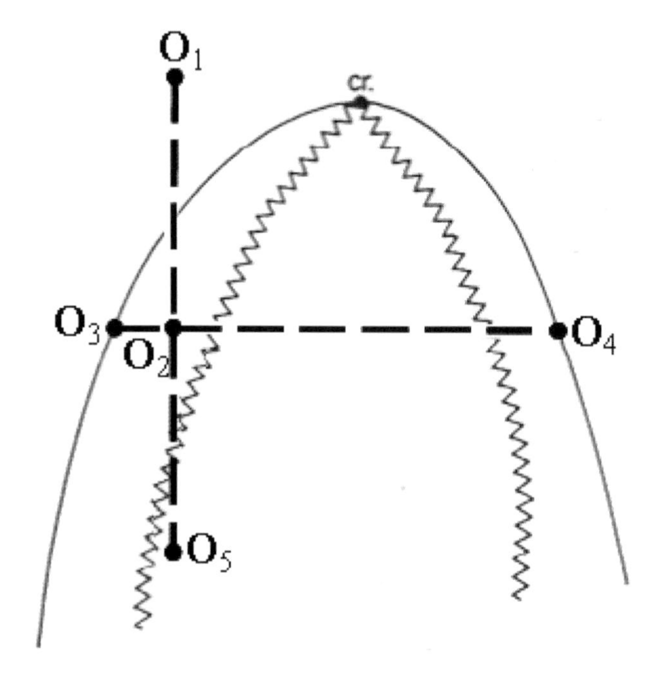

Fig. 4.1 The temperature-concentration phase diagram of a binary mixture. Quenching is performed from the original homogeneous state O_1 to the metastable state O_2 or to the non-stable state O_5. In the former case, phase separation to the final stable state, described by points O_3 and O_4, occurs through the nucleation process, whereas in the latter case, through spinodal decomposition. Figure 4.1 describes the upper critical solution point. For the lower critical solution point, the temperature scale may be inverted.

corresponding to point O_3. Then, a chemical reaction leads to an additional driving force $A \neq 0$, shifting the system homogeneously from a quenched metastable state to the closest stable state O_3. The points O_3 and O_4 are separated by a singular line defining the stability boundary. Therefore, spatially homogeneous relaxation from O_2 to O_4 is impossible. Strictly speaking, for such reactive systems, the first stage of nucleation does not exist. Immediately after a quench, the system is shifted by chemistry to the closest state on the coexistence curve at the same temperature. The second possibility takes place when the forward and backward reactions rates are equal. If, in addition, the rate of reaction is much faster than that of phase separation, the chemical reaction, which tends to randomly mix the mixture, will keep the mixture homogeneous.

These physical arguments are supported by simple model calculations. Let us start with the simplest model of a strictly regular solution [114], where the chemical potentials have the following form [3]

$$\mu_1 = \mu_1^0 + \kappa_B T \ln x + \omega (1-x)^2,$$
$$\mu_2 = \mu_2^0 + \kappa_B T \ln (1-x) + \omega x^2. \tag{4.6}$$

The coexistence curve is defined by Eq. (4.5). Using (4.6) gives $x \equiv x' = 1 - x''$ for the concentrations of the two coexistence phases [3], following the symmetric coexistence curve,

$$\frac{\omega}{\kappa_B T} = \frac{\ln x - \ln (1-x)}{2x-1}. \tag{4.7}$$

Using (4.4), the law of mass action can be written as

$$\frac{\omega}{\kappa_B T} = \frac{\ln(1-x) - \delta \ln x}{\Lambda + \delta (1-x)^2 - x^2} \tag{4.8}$$

where

$$\delta = -\frac{\nu_1}{\nu_2}; \qquad \Lambda = \frac{\delta \mu_1^0 - \mu_2^0}{\omega}. \tag{4.9}$$

Phase separation in a reactive binary mixture will occur if and only if Eqs. (4.7) and (4.8) have a common solution. One can immediately see that Eqs. (4.7) and (4.8) coincide if both $\Lambda = 0$ ("symmetric" mixture) and $\delta = 1$ (isomerization reaction). If $\delta = 1$, $\Lambda \neq 0$, these equations have no common solution, and the strictly regular reactive binary mixture will not separate into two phases. In all other cases, one can equate the right-hand sides of Eqs. (4.7) and (4.8) to obtain

$$\frac{(1-x)^2 \ln (1-x) - x^2 \ln x}{\ln x - \ln (1-x)} = \frac{\Lambda}{\delta - 1} \equiv q. \tag{4.10}$$

The concentration x ranges from zero to unity. Therefore, Eq. (4.10) has a solution only for $q < (2 \ln 2 - 1)/4 = 0.097$. The model of a strictly regular reactive binary mixture allows phase separation only for substances and chemical reactions satisfying this inequality. If, for example, $\Lambda = 0.09$ and $\delta = 2$, the criterion $q < 0.097$ is satisfied, two phases may coexist, and their concentrations are given by intersection of the coexistence curve (4.7) and the law of mass action (4.8). By contrast, for $\Lambda = 0.1$ and $\delta = 2$, this criterion is not satisfied, and hence there is no phase separation.

The non-compatibility of Eqs. (4.7) and (4.8) for some range of temperatures and pressures, i.e., the nonexistence of a common solution for the concentrations in the range from zero to unity, may appear in some models of an n-component system with $n-1$ independent chemical reactions. In all other cases, the existence of r chemical reactions will decrease by r the dimension of the coexistence hypersurface.

Phase separation in binary mixtures has been described by models more complex than that of the strictly regular solution, such as the van Laar and Margules models [115].

The calculations performed above for a chemical reaction in a strictly regular binary mixture can easily be generalized to a three-component mixture of components $A + B \rightleftarrows C$, described by the activity coefficients

$$\ln \gamma_1 = \omega x_2 (1 - x_1); \ \ln \gamma_2 = \omega x_1 (1 - x_2); \ \ln \gamma_3 = -\omega x_1 x_2. \tag{4.11}$$

If there are two independent chemical reactions, a system can only exist in one phase. However, when there is a single reaction, the situation is very different. A complete analysis has been given for a reversible bimolecular reaction that involves all three species [116].

4.2.2 *Thermodynamic analysis of reactive ternary mixtures*

Another way to find the global phase diagram for reactive ternary mixtures is by topological analysis of the intersection of the critical manifolds of the model considered and the specified chemical equilibrium surface. Let us start with the mean-field model described by Eq. (4.1) with activity coefficients given by (4.11) [117]. Phase equilibrium is obtained by demanding that all three chemical potentials be equal in coexistence phases. The coexistence curve satisfies the following symmetry relations: $x_1' = x_1'$; $x_2' = x_2'$; $x_3' = x_3'$.

The phase equilibrium surface has the following form,

$$\frac{\omega}{\kappa_B T} = \frac{1}{x_1 - x_2} \ln \frac{x_1}{x_2} \tag{4.12}$$

and the critical curve is given by the limit $x_1 - x_2 \to 0$,

$$x_{1,cr} = x_{2,cr} = \frac{\kappa_B T_{cr}}{\omega}. \tag{4.13}$$

The constraint $A = \mu_1 - \mu_2 - \mu_3 = 0$, imposed by the chemical reaction, takes the following form

$$x_3 = \exp\left(-\frac{\Delta G_0}{\kappa_B T}\right) x_1 x_2 \exp\left[\frac{\omega (x_1 + x_2 - x_1 x_2)}{\kappa_B T}\right] \tag{4.14}$$

where ΔG_0 is the deviation of the standard Gibbs free energy from a hypothetical ideal solution state, which can be written as $\Delta G_0 = \Delta H_0 - T\Delta S_0$, with ΔH_0 and ΔS_0 being the standard enthalpy and entropy for the given reaction. The intersection of this chemical equilibrium surface with the coexistence surface (4.12) defines a unique coexistence curve in the plane $(\kappa_B T/\omega,\, x_1)$ for fixed values of $\Delta G_0/\kappa_B T$. Numerical calculations for specific values of $\Delta G_0/\ \kappa_B T$ show [117] that $A - B$ repulsion results in phase separation as the temperature is lowered. Upon further lowering of the temperature, the formation of C becomes favorable, thereby reducing the unfavorable mixing, and the phases become miscible again. At low enough temperatures, the solution consists only of C and either A or B. Hence, the presence of a chemical reaction might result in the appearance of the lower critical consolute point, which, together with an existing upper critical consolute point, results in a close-loop coexistence curve. We will consider this problem in Sec. 5.4.

Another peculiarity of the phase diagrams in reactive ternary mixtures has been found [118] in the context of a slightly more complex model for $A + B \rightleftarrows C$ reaction, in which the chemical potentials of the components have the following form,

$$\begin{aligned} \mu_1 &= \mu_1^0 + \kappa_B T \ln x_1 + b x_3 + c x_2 - W, \\ \mu_2 &= \mu_2^0 + \kappa_B T \ln x_2 + a x_3 + c x_1 - W, \\ \mu_3 &= \mu_3^0 + \kappa_B T \ln x_3 + a x_2 + b x_1 - W \end{aligned} \tag{4.15}$$

where $W = a x_2 x_3 + b x_1 x_3 + c x_1 x_2$. The chemical equilibrium constraint $A = 0$ for Eqs. (4.15) can be expressed as follows,

$$\begin{aligned} \kappa_B T \ln \frac{x_3}{x_1 x_2} &= a\,(x_3 - x_2) + b\,(x_3 - x_1) \\ &\quad + c\,(x_1 + x_2) - W - \Delta H_0 + T\Delta S_0. \end{aligned} \tag{4.16}$$

The intersection of this surface and the critical manifolds following from (4.15) gives the phase diagram of a reactive ternary mixture. The results depend on whether the chemical process is enthalpically or entropically favored. Detailed analysis shows [118] that there are six different types of phase diagrams, depending on the combinations of sign of the parameters a, b and c which include the triple and quadruple points, closed-loop phase coexistance curves, azeotrope-like points, etc.

As an example of homogeneous nucleation in a chemically reactive system, let us return [119] to the simple bistable chemical reaction (the second

Schlogl trimolecular model) considered in Sec. 1.3

$$A + X \underset{k_2}{\overset{k_1}{\rightleftarrows}} 3X; \qquad X \underset{k_4}{\overset{k_3}{\rightleftarrows}} B \tag{4.17}$$

with the concentrations x_A and x_B of species A and B being held constant. The kinetic equation for the concentration x of the component X is

$$\frac{dx}{dt} = -k_1 x^3 + k_2 x_A x^2 - k_3 x + k_4 x_B \equiv -\frac{dV(x)}{dx}. \tag{4.18}$$

Equation (4.18) has three solutions. We assume that the values of the coefficients k_i provide two stable solutions x_1 and x_3, and one unstable solution x_2 such that $x_1 < x_2 < x_3$. Moreover, it is suggested that $V(x_1) < V(x_3)$, i.e., the state x_3 is stable while the state x_1 is metastable. If the reaction (4.17), supplemented by diffusion, occurs in a closed one-dimensional vessel, Eq. (4.18) will have the form of the reaction-diffusion equation

$$\frac{\partial x(\rho, t)}{\partial t} = -k_1 x^3 + k_2 x_A x^2 - k_3 x + k_4 x_B + \mathcal{D}\frac{\partial^2 x}{\partial \rho^2} \tag{4.19}$$

where ρ is the spatial coordinate and $\mathcal{D}$ is the diffusion coefficient. Introducing the dimensionless variables $u_i = x_i/x_1$; $\tau = tk_1 x_1^2$; $D = \mathcal{D}/k_1 x_1^2$, and moving the origin to the metastable state x_1, $\zeta = (x - x_1)/x_1$, one gets

$$\frac{\partial \zeta}{\partial \tau} = D\frac{\partial^2 \zeta}{\partial \rho^2} - b\zeta + a\zeta^2 - \zeta^3 \tag{4.20}$$

with $b = (u_2 - 1)(u_3 - 1)$ and $a = u_2 + u_3 - 2$. If the system is quenched to the metastable state $\zeta = 0$, its transition to the stable state $\zeta = \zeta_3$, can be understood as the spontaneous creation of two solutions of Eq. (4.20) ("kink"-"antikink")

$$\zeta(\rho, t) = \zeta_3 \left\{1 + \exp\left[\pm\delta(\rho - \rho_0 \pm vt)\right]\right\}^{-1} \tag{4.21}$$

where

$$\zeta_3 = \frac{1}{2}\left[a + \left(a^2 - 4b\right)^{1/2}\right]; \; \delta^2 = \frac{\zeta_3^2}{2D}; \; v = -\frac{3b - a\zeta_3}{2\delta}. \tag{4.22}$$

The kink-antikink pair separates with time in opposite directions. When separated by a certain critical length κ, equal to $-\delta^{-1}\ln\left[\left(a - 3\sqrt{b/2}\right)\left(a + 3\sqrt{b/2}\right)^{-1}\right]$, they create a stationary profile, called a nucleation nucleus $\zeta(\rho)$. When the kink and the antikink are separated by a distance less than κ, they annihilate, and the concentration profile returns to the homogeneous metastable state. When the separation becomes greater than κ, the kink-antikink pair separates, and the system evolves toward the homogeneous stable state. Equation (4.21) becomes simplified in the small-amplitude-nucleus limit, when $\kappa \to 0$, and the nucleation nucleus then becomes

$$\zeta(\rho) = \frac{3b}{a}\left[1 + \cosh(\delta\rho)\right]^{-1} = \frac{3b}{2a}\,\mathrm{sech}^2\left(\frac{\delta\rho}{2}\right). \tag{4.23}$$

Thus far, we have discussed a deterministic system. The influence of homogeneous external white noise and internal chemical fluctuations has been considered [119] in the framework of the extended Kramers approach and the multivariate master equation, respectively. The nucleation due to the internal fluctuations is limited to a very narrow region near the transition, and it is difficult to observe this experimentally. The external noise may be easily greater than the internal noise and, therefore, more important. In addition to the homogeneous nucleation initiated by the fluctuations (considered above), the important problem is non-homogeneous nucleation initiated, say, by pieces of dust present in solutions.

4.2.3 *Kinetics of phase separation*

Let us first discuss non-reactive systems. All thermodynamic states within the coexistence curve shown by the solid line in Fig. 4.1 belong to two-phase states. Therefore, after a change in temperature (or pressure) from point O_1 to O_2, phase separation will occur and the system will consist of two phases described by the thermodynamic condition (4.5) and shown by points O_3 and O_4 in the figure. All states located within the coexistence curve and close to it are the so-called metastable states. Although the energy of these states is larger than that of the corresponding two-phase system, one needs a finite "push" to pass from the local minimum in energy to the deeper global minimum. Such a "push" is provided by the thermal fluctuations in density for a one-component system and in concentration for many-component mixtures. Therefore, these fluctuations are of crucial importance for the analysis of the decay of a metastable state.

The change of the thermodynamic potential associated with the appearance of the nucleus of the new phase (assumed, for simplicity, to be spherical with radius r) is equal to

$$\Delta G(r) = -\Delta\mu\frac{4}{3}\pi r^3 + 4\pi\sigma r^2 \tag{4.24}$$

where $\Delta\mu$ is the energy gain in a stable state compared to a metastable state, and σ is the surface tension. The first (negative) term in (4.24) encourages phase separation while the second (positive) term prevents it. It is clear from (4.24) that for small nuclei, $\Delta G > 0$. That is, small nuclei tend to shrink because of their high surface-to-volume ratio. Only nuclei larger than the critical size r_C are energetically favorable. From $\Delta G/dr = 0$, one finds

$$r_C = \frac{2\sigma}{\Delta\mu}. \tag{4.25}$$

The traditional phenomenological approach to the decay of a metastable state of a pure substance is based on the distribution function $W(r,t)$ of nuclei of size r at time t. The continuity equation has the form,

$$\frac{\partial W}{\partial t} = \frac{\partial}{\partial r}\left[FW + D\frac{\partial W}{\partial r}\right] \equiv -\frac{\partial J}{\partial t} \tag{4.26}$$

where the flux $J(r,t)$ of nuclei along the size axis is determined by two unknown functions F and W. One can reduce [120] the equation(s) of the critical dynamics to the Langevin equation for the radius of nuclei with a known random force, which can be transformed to the Fokker-Planck equation (4.26) with functions F and D of the form

$$F(r) = D_0\left(\frac{1}{r} - \frac{1}{r_C}\right); \qquad D(r) = \frac{D_0\kappa_B T}{8\pi\sigma r^2} \tag{4.27}$$

where D_0 is the diffusion coefficient far from the critical point.

The nucleation process in a binary mixture can be described in analogous fashion [121], [18]. The formation energy of a nucleus containing n_1 atoms of component 1 and n_2 atoms of component 2 is

$$\Delta G = (\mu_{1,n} - \mu_1)\, n_1 + (\mu_{2,n} - \mu_2)\, n_2 + 4\pi\sigma r^2 \tag{4.28}$$

where μ_1 and μ_2 are the chemical potentials of the components in the homogeneous phase, and $\mu_{1,n}$ and $\mu_{2,n}$ are the corresponding quantities in the nuclei. The size r of a nucleus is related to its structure by $4\pi r^3/3 = v_1 n_1 + v_2 n_2$, where v_1 and v_2 are the volumes per atom.

In contrast to a pure substance, the critical nucleus is defined not only by its size r, but also by the concentration $x = n_1/(n_1 + n_2)$. Therefore, the height of the potential barrier is defined by $\partial\Delta G/\partial r = \partial\Delta G/\partial x = 0$, which gives the following relation between r_C and x_C of the critical nucleus,

$$r_C = 2\sigma\frac{x_C v_1 + (1 - x_C)\, v_2}{x_C\Delta\mu_1 + (1 - x_C)\,\Delta\mu_2} \tag{4.29}$$

where $\Delta\mu_i = \mu_{i,n} - \mu_i$.

Let us now consider reactive systems. The concentration x changes as a result of a chemical reaction, which shifts the initial quenched state to the closest equilibrium state on the coexistence curve. Each intermediate state corresponds to a different radius of critical nucleus $r_C(t)$, increasing toward the coexistence curve, where $r_C \to \infty$. Hence, one has to replace r_C by $r_C(t)$ everywhere. The latter function can be found from Eq. (4.29) under the assumption of a quasi-static chemical shift. This shift is caused by a chemical reaction, and can be described in the linear approximation by

$$\frac{dx}{dt} = -\frac{x - x'}{\tau} \tag{4.30}$$

where τ^{-1} is the rate of the chemical reaction.

Inserting the solution of Eq. (4.30), $x - x' = (x_0 - x')\exp\left(-\frac{t}{\tau}\right)$, into Eq. (4.29), one obtains

$$\begin{aligned} r_C(t) &= \frac{2\sigma\left[x''v_1 + (1 - x'')v_2\right]}{\kappa_B T\left[x''/x' - (1 - x'')/(1 - x')\right](x' - x'')} \\ &\equiv r_0 \exp(t/\tau) \end{aligned} \tag{4.31}$$

where we used the coexistence condition (4.5), and retained only the leading logarithmic part of the chemical potentials.

The next step is to obtain the Fokker-Planck equation for the distribution function $W(r, t, x)$ for nuclei of size r and composition x at time t. We make the approximation that all nuclei which are important for phase separation have the same composition x''. Therefore, we may omit the argument x in $W(r, t, x)$. In other words, one assumes that a chemical reaction brings the path leading to the saddle point closer to that of $x = x''$.

Substituting (4.27) and (4.31) into the Fokker-Planck equation (4.26), one obtains

$$\frac{\partial W}{\partial t} = \frac{\partial}{\partial r}\left\{D_0\left[\frac{1}{r} - \frac{1}{r_0}\exp\left(-\frac{t}{\tau}\right)\right]W + \frac{D_0\kappa_B T}{8\pi\sigma r^2}\frac{\partial W}{\partial r}\right\}. \tag{4.32}$$

To find the approximate solution of this complicated equation, we first consider a simplified version, which allows an exact solution. One replaces functions $F(r)$ and $D(r)$ in (4.27) by simpler functions which incorporate the main property of (4.27). According to the physical picture, $F(r)$ is positive for $r < r_C$ and negative for $r > r_C$. Therefore, we approximate $F(r)$ by

$$F(r) = B(r_C - r). \tag{4.33}$$

After choosing this form of $F(r)$, we are no longer free to choose the second function $D(r)$. These two functions are not independent because there is no flux in equilibrium, which, according to (4.26), leads to $W_{eq} = \exp\left[-\int drF(r)/D(r)\right]$. On the other hand, by definition, $W = \exp(-\Delta G/k_BT)$, and, therefore,

$$\int \frac{F(r)}{D(r)}dr = \frac{\Delta G}{k_BT} \tag{4.34}$$

where $\Delta G(r)$ and r_C are given by (4.24) and (4.25). A simple calculation yields

$$D(r) = \frac{Bk_BTr_C}{8\pi\sigma r} \approx \frac{Bk_BT}{8\pi\sigma}. \tag{4.35}$$

We keep only the largest term in $r - r_C$ in the last equality in (4.35). Using (4.33), (4.31) and (4.35), one obtains the simplified Fokker-Planck equation,

$$\frac{\partial W}{\partial t} = \frac{\partial}{\partial r}\left[B\left(r_0\exp\left(\frac{t}{\tau}\right) - r\right)W\right] + D_0\frac{\partial^2 W}{\partial r^2}, \tag{4.36}$$

where $D_0 = Bk_BT/8\pi\sigma$. Equation (4.36) can be solved exactly by introducing the characteristic function

$$W(K,t) = \int W(r,t)\exp(iKr)\,dr. \tag{4.37}$$

The original Fokker-Planck equation (4.26)–(4.27) is thus transformed into a first-order linear partial differential equation for $W(K,t)$ [122],

$$\frac{\partial W}{\partial t} - BK\frac{\partial W}{\partial K} = -\left[D_0K^2 + iKBr_0\exp(\frac{t}{\tau})\right]W. \tag{4.38}$$

To solve this equation, one uses the method of characteristics, with initial condition

$$W(r,t=0) = \delta(r-\xi) \tag{4.39}$$

where ξ is of order of a single molecule (there are no nuclei immediately after a quench). Solving Eq. (4.38) and performing the inverse Fourier transformation, one obtains the Gaussian distribution function

$$W(r,t) = \frac{1}{\sqrt{4\pi\overline{\Delta r^2}}}\exp\left[-\frac{(r-\overline{r})^2}{2\overline{\Delta r^2}}\right] \tag{4.40}$$

with mean value $\overline{r}$ and variance $\overline{\Delta r^2}$ given by

$$\overline{r} = \xi\exp(Bt) - \frac{Br_0}{\tau^{-1} - B}\left[\exp\left(\frac{t}{\tau}\right) - \exp(Bt)\right],$$

$$\overline{\Delta r^2} = \frac{D_0}{2B}\left[\exp(2Bt) - 1\right]. \tag{4.41}$$

Two characteristic times, B^{-1} and τ, are associated with transient processes and chemical reactions, respectively. However, the solution (4.40) diverges with time, so that Eq. (4.36) does not have a steady-state solution. Hence, instead of the exact solution, one has to use the approximate solution.

One can find [122] the appropriate solution of Eq. (4.36), but we prefer to find the solution of the original equations (4.26) and (4.27), assuming that B^{-1} is very small [123]. This assumption permits us to neglect transient processes, and, for $B^{-1} < \tau$, restrict attention to those solutions of (4.36) which supply the quasi-static flux $J_{qss}(t)$ independent on the size of the nuclei. We consider the quasi-steady-state regime which is established after the transients disappear. The quasi-steady-state solution $W_{qss}[r, r_C(t)]$ does not have an explicit time dependence, so that $\partial W_{qss}/\partial t = 0$, and the Fokker-Planck equation (4.32) can be rewritten as

$$D_0\left[\frac{1}{r} - \frac{1}{r_0}\exp\left(-\frac{t}{\tau}\right)\right] W_{qss} + \frac{D_0 \kappa_B T}{8\pi\sigma r^2}\frac{\partial W_{qss}}{\partial r} \equiv J_{qss}\left(\frac{t}{\tau}\right). \tag{4.42}$$

The quasi-steady-state flux J_{qss} reduces to the steady-state flux J_{ss} when the chemical reaction is absent,

$$J_{qss}\left(\frac{t}{\tau} = 0\right) = J_{ss}. \tag{4.43}$$

There is no stationary state for the reactive system considered here. Therefore, one is forced to give a new definition to the lifetime of a metastable state in reactive systems. The simplest generalization is the time required to produce one critical nucleus

$$\int_0^{T_{ch}} dt\ J_{qss}(t) = 1. \tag{4.44}$$

For the time-independent case, J_{qss} is replaced by J_{ss}, according to (4.43), and T_{ch} is replaced by the lifetime of the metastable state in the non-reactive system, $T_0 = J_{ss}^{-1}$.

Let us now turn to the solution of Eq. (4.42). The boundary conditions for this equation are determined by the requirements that the distribution of nuclei of minimal size ξ will be the equilibrium distribution, and the total number of nuclei in the system is bounded,

$$W_{qss}(\xi) = W_{eq}(\xi); \qquad W_{qss}(r \to \infty) = 0 \tag{4.45}$$

where W_{eq} corresponds to zero flux.

The solution of Eq. (4.42) which satisfies the boundary conditions (4.45) has the form [123],

$$W_{qss} = W_{eq}(r)\left[1 - J_{qss}\int_{\xi}^{r}\frac{dz}{W_{eq}(z)\,D(z)}\right] \tag{4.46}$$

where

$$J_{qss}^{-1} = \int_{\xi}^{\infty}\frac{dz}{W_{eq}(z)\;D(z)}. \tag{4.47}$$

The function $W_{eq}^{-1}(r)$ has a sharp maximum at r_C, which reflects the existence of a barrier to nucleation. Therefore, the integral (4.47) can be evaluated by the method of steepest descents,

$$J_{qss} = \sqrt{\frac{\kappa_B T}{4\pi^2\sigma}\frac{D_0}{2r_0^2}}\exp\left\{-\frac{2t}{\tau} - \frac{\Delta G(r_0)}{\kappa_B T}\exp\left(\frac{2t}{\tau}\right)\right\} \tag{4.48}$$

where $\Delta G(r_0) = 4\pi\sigma r_0^3/3$ is the minimal work needed to produce the critical nucleus in the initial state immediately after a quench.

Upon inserting J_{ss}, defined by (4.43), into (4.48), one obtains

$$J_{qss} = J_{ss}\exp\left\{-\frac{2t}{\tau} - u\left[\exp\left(\frac{2t}{\tau}\right) - 1\right]\right\}, \tag{4.49}$$

where $u = \Delta G(r_0)/\kappa_B T$.

Finally, inserting (4.49) and $T_0 = J_{ss}^{-1}$ into (4.44) gives the equation for the lifetime T_{ch} of a metastable state in a reactive system as a function of the lifetime in the non-reactive system T_0 (the latter depends on the volume under observation), the extent of quench u, and the rate τ^{-1} of the chemical reaction,

$$1 = \frac{\tau u\exp(u)}{2T_0}\int_{u}^{u\exp(2T_{ch}/\tau)}\frac{\exp(-z)}{z^2}dz. \tag{4.50}$$

Equation (4.50) for T_{ch} has been solved numerically. This equation has no solution if τ is very small, i.e., the chemical reaction is very fast. Then, although the thermodynamics allows phase separation in a reactive system, such a separation is impossible from the kinetic point of view. The system is dragged by the chemical reaction to a homogeneous equilibrium state on the coexistence curve before the nuclei of the new phase appear.

The minimal τ which allows phase separation for different quenches (with characteristic u and T_0) is given approximately by the following formula,

$$\left(\frac{\tau}{T_0}\right)_{\min} \approx 3.03 + 2.08u. \tag{4.51}$$

For smaller values of τ, the system will never separate into two phases. The two possible sets of experiments performed with and without a catalyst (different quenches for a given chemical reaction and different chemical reactions for the same quench) will be discussed in the next section.

4.3 Phase equilibrium in reactive mixtures quenched into an unstable state

Thus far, we have considered phase separation in a system quenched into the metastable region of phase diagram (point O_2 in Fig. 4.1) located between the limit of thermodynamic stability (spinodal) and the boundaries of phase coexistence (binodal). In that case, separation proceeds through nucleation. Another mechanism for phase separation takes place when a system is quenched into the unstable region (point O_5 in Fig. 4.1) of the phase diagram located inside the spinodal curve. In this case, the process leads to macroscopic phase separation via a complicated labyrinthine interconnected morphology. The phenomenological description of this process is based on the Cahn-Hillard equation [124] which, for our case of the conserved variable, has the following form

$$\frac{\partial \Psi}{\partial t} = \Gamma \nabla^2 \frac{\partial F[\Psi]}{\partial \Psi} \tag{4.52}$$

where $\Psi = x_1 - x_2$ is the concentration difference (order parameter), Γ is proportional to the mobility, and $F[\Psi]$ is the Ginzburg-Landau free energy functional

$$F[\Psi] = \int dx \left[\kappa \mid \nabla \Psi \mid^2 + f(\Psi) \right], \tag{4.53}$$

where $f(\Psi)$ is the double-well potential, $f(\Psi) = -a\Psi^2 + b\Psi^4$, and κ is the positive constant related to the interaction range. When the $A-B$ mixture is quenched from a homogeneous state (O_1 in Fig. 4.1) to the non-stable state O_5, the system separates into A-rich and B-rich phases. Numerical simulation has been performed [125] by placing Eq. (4.52) on a 128×128 square lattice with periodic boundary conditions, with the lattice site and time step both set at unity, and the local part of free energy of the form

$$\frac{df(\Psi)}{d\Psi} = -1.3 \tanh \Psi + \Psi. \tag{4.54}$$

The order parameter $\Psi = \Psi_0 + \delta\Psi$ at each site is given by $\Psi_0 = 0.4$, and the random number $\delta\Psi$ chosen in the range $(-0.15, 0.15)$. For an A-rich mixture (with concentrations of A and B being $x_1 = 0.7$ and $x_2 = 0.3$, respectively), the B-rich domains growing in the A-rich matrix are shown in Fig. 4.2.

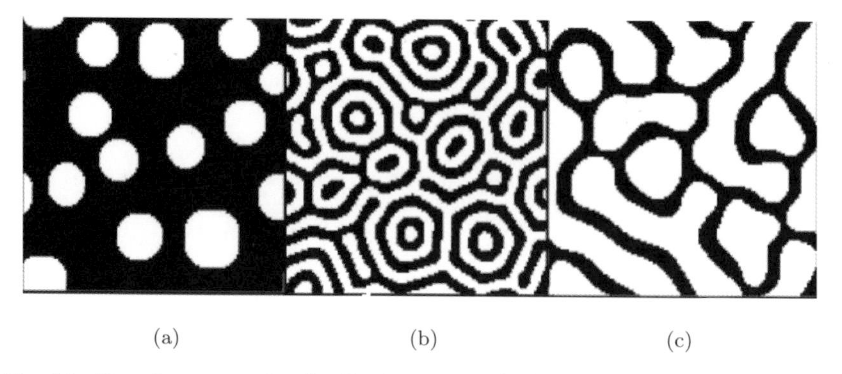

Fig. 4.2 Domain patterns for $A - B$ mixture quenched to a non-stable state. From left to right: (a) after 10^5 steps with no chemical reaction; (b) after $t = 1.32 \times 10^5$ steps with the chemical reaction ($k_f = k_b = 0.01$) started at $t = 10^5$; (c) after $t = 8.2 \times 10^4$ steps with the chemical reaction ($k_f = 0.0013$ and $k_b = 0.0007$) started at $t = 5 \times 10^4$. Reproduced from Ref. [125] with permission, copyright (1997), Physical Society of Japan.

The morphologies of phase separation become much more manifest in the presence of an ongoing chemical reaction. Unlike the previously considered phase separation from a metastable state in a closed system, we now assume that our system is open, i.e., far from equilibrium, so that the chemical reaction is maintained by an external agent, say, light (for a discussion of this point, see [126], [127]). The simplest way to take into account the chemical reaction $A \leftrightarrow B$ is to add to Eq. (4.52) the equation of the chemical reaction, $k_f x_1 - k_b x_2$, where k_f and k_b are the forward and backward reaction rates. Thus, we rewrite Eq. (4.52) for the reactive binary mixture in the following form [128],

$$\begin{aligned} \frac{\partial \Psi}{\partial t} &= \Gamma \nabla^2 \frac{\delta F[\Psi]}{\delta \Psi} - k_f \Psi + k_b (1 - \Psi) \\ &= \Gamma \nabla^2 \left(\frac{\partial f}{\partial \Psi} - 2\kappa \nabla^2 \Psi \right) - (k_f + k_b) \Psi + k_b \,. \end{aligned} \tag{4.55}$$

If the chemical reaction starts at $t = 10^5$ with $k_f = k_b = 0.01$, the domain patterns shown in Fig. 4.2a gradually change to the concentric patterns shown in Fig. 4.2b for $t = 1.32 \times 10^5$ [125]. The difference between the patterns in Figs. 4.2a and 4.2b are connected with the time delay between the beginning of phase separation after the temperature quench and the onset of the chemical reaction. If phase separation begins simultaneously with the chemical reaction, lamellar patterns will form. Still other patterns are formed when the rates of the forward and back reactions are different, $k_f \neq k_b$. In the system shown in Fig. 4.2c, the non-symmetrical chemical reaction ($k_f = 0.0013$ and $k_b = 0.007$) started at $t = 5 \times 10^4$ steps after the quench. As a result, a bicontinuous pattern emerges at $t = 8.2 \times 10^4$ steps [125]. In this case, even without a chemical reaction, one expects A -rich droplets in the B-rich phase because $k_b > k_f$.

The increase in phase separation coupled with a chemical reaction has been calculated [129] by including hydrodynamic effects. Equation (4.55) has been coupled with the Navier-Stokes equation. Numerical simulations have been performed [129] under conditions similar to those described above (256×256 square lattice, $\Psi_0 = 0.1$ and $\delta\Psi = (-0.025, 0.025)$). The influence of hydrodynamic effects is characterized by the viscosity η. When the viscosity is low ($\eta = 1$), hydrodynamics plays an important role, whereas for low fluidity ($\eta = 100$), hydrodynamic effects are not important.

Domain patterns at $t = 10^4$ are shown in Figs. 4.3–4.4 for $\eta = 1$ (left row), and for $\eta = 100$ (right row). Figures 4.3a–4.3c describe the symmetric reactions (from the top down, $k_f = k_b$ equals 10^{-4}, 10^{-3} and 10^{-2}), while Figs. 4.4a–4.4c show the results for non-symmetric reactions (from the top down: $k_f = 1.1 \times 10^{-4}$; 1.1×10^{-3}; 1.1×10^{-2} and $k_b = 9 \times 10^{-5}$; 9×10^{-4}; 9×10^{-3}). These figures clearly show the competition between hydrodynamics effects and chemical reactions. As seen in Fig. 4.3a, for a low reaction rate, isolated domains become more circular under hydrodynamic flow. For intermediate reaction rates (Fig. 4.3b), hydrodynamic effects lead to a circular shape for isolated domains since the chemical reaction rate is not high enough to make discontinuous domains penetrate each other. The latter effect occurs at higher reaction rates (Fig. 4.3c), where the mixing process due to the chemical reaction proceeds more quickly than the thermodynamically induced phase separation. The results are quite similar for non-symmetric reactions ($k_f \neq k_b$), as shown in Fig. 4.4. Some difference occurs for intermediate reaction rates (Fig. 4.4b), where a continuous pattern emerges along with an isolated A -rich phase for $\eta = 100$. However, for $\eta = 1$, the disconnected A -rich domain appears in the matrix of the B-rich phase, and the B-rich phases do not penetrate each other.

The additive contribution of the two ingredients of the process — segregation dynamics and chemical reaction — can be demonstrated by starting

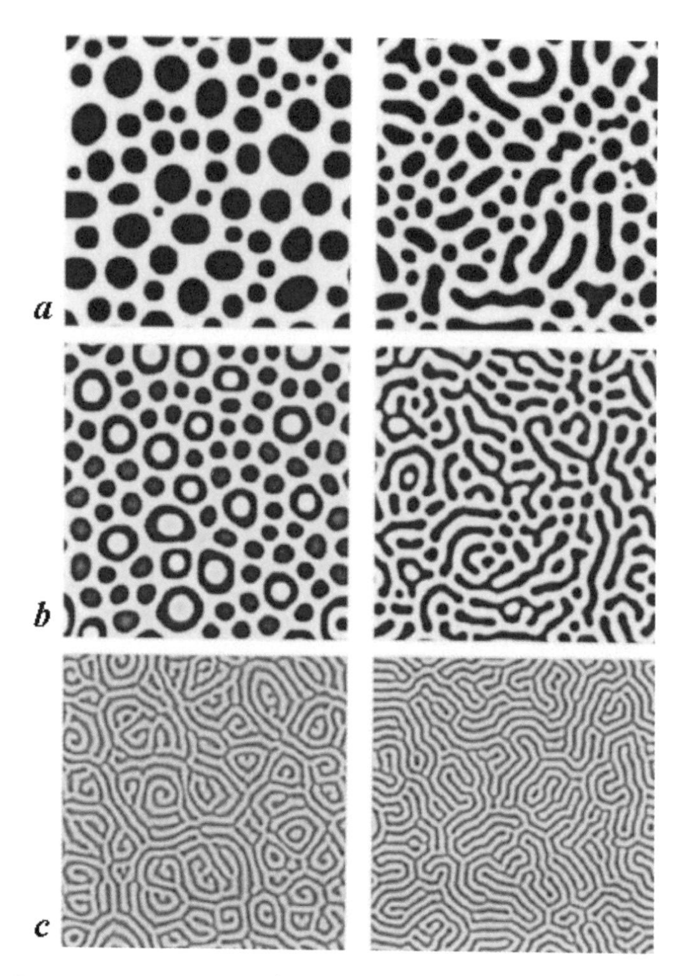

Fig. 4.3 Domain patterns at $t = 10^4$ steps with and without hydrodynamic effects for symmetric chemical reactions. A-rich and B-rich regions are shown as white and black, respectively (see text for explanation). Reproduced from Ref. [129] with permission, copyright (2003), American Institute of Physics.

from a different model [130], in which the dynamics of a quenched system is described in the context of an Ising model with a spin variable σ_i at site i. The two states of the spin-1/2 Ising model correspond to the site occupied by atom 1 or 2. The dynamics of the system are described by spin-exchange or Kawasaki dynamics, whereas the chemical contribution is described by a master equation for the time-dependent probability distribution for N

a

b

c

Fig. 4.4 Domain patterns at $t = 10^4$ steps with and without hydrodynamic effects for non-symmetric chemical reactions. A-rich and B-rich regions are shown as white and black, respectively (see text for explanation). Reproduced from Ref. [129] with permission, copyright (2003), American Institute of Physics

spins, which, after averaging over all configurations, leads to Eq. (4.55) [130]. Note that the two last "chemical" terms in Eq. (4.55) can be added [131] to the free energy functional $F[\Psi]$ in the following form

$$F[\Psi] \to F[\Psi] + \tag{4.56}$$

$$+\frac{k_f + k_b}{2} \int\int dz dz' G(z, z') \left[\Psi(z,t) - \overline{\Psi}\right] \left[\Psi(z',t) - \overline{\Psi}\right]$$

where $\overline{\Psi} = (k_f - k_b) / (k_f + k_b)$ and $G(z, z')$ is the Green function for the Laplace equation $\nabla^2 G(z, z') = -\delta(z - z')$ with appropriate boundary conditions, so that for $d = 3$, $G(z, z') = 4\pi |z - z'|^{-1}$. Including the chemical reaction is thus equivalent to including a long-range repulsive interaction, thereby transforming the problem of a reactive mixture to that of a non-reactive mixture with competing short-range and long-range interactions.

A linear stability analysis of Eq. (4.55) in the absence of chemical reactions predicts exponential growth of the concentration fluctuations with a growth factor that has a cutoff at large wavenumber k_C (small wavelength). Thus, fluctuations with $k < k_C$ grow while those with $k > k_C$ decay. It was shown [128] that the occurrence of a reaction results in shifting the cutoff by an amount proportional to $k_f + k_b$, i.e., the presence of a chemical reaction makes a system more stable with respect to phase separation. This result of linear stability analysis was verified by numerical simulations of the full nonlinear equation [128] , as well as for the Kawasaki exchange interaction model including the chemical reaction [132].

Other interesting results obtained for both models relate to the growth of the steady-state domains of size R_{ss}. For the chemical reaction with $k_f = k_b \equiv k$, $R_{ss} \sim k^{-\phi}$ has the same critical index $\phi = 0.3$ as the growth index $(R(t) \sim t^{\phi})$ in the absence of a chemical reaction. The latter result has been refined [133]. A more serious discrepancy was found [134] when, in contrast to the numerical calculation performed using the Monte Carlo technique for particles on a square lattice [132], the calculations were carried out in continuous space using molecular dynamics. Although these two numerical techniques usually give the same results for equilibrium properties, the results are different for non-equilibrium dynamics, where the collective hydrodynamics modes are not simulated in the typical single-particle Monte-Carlo method. Thus, no correction was found for the growth exponent in the absence of a chemical reaction and scales with respect to the reaction rate in reactive systems.

The contributions of critical dynamics (diffusion) and the chemical reaction were previously considered independently. Moreover, contrary to critical dynamics, the chemical rate laws do not take into account the non-ideality of the systems. Such an approach is inconsistent [135] because it is characteristic of chemical reactions and diffusion that the thermodynamic forces governing these processes are related, with both being functions of the chemical potentials (in the former case, of the spatial derivatives of chemical potentials). It is for this reason that for equilibrium systems, diffusional stability automatically insures the stability of chemical equilibrium (Duhem-Joguet theorem [3]). A consistent analysis shows [135] that the interesting and possible important idea of applications to the freezing of the phase separation by the use of simultaneous chemical reactions, is not easily carried out in practice and demands special requirements both for the systems and for the chemical reactions.

Thus far, we have considered the influence of the $A \rightleftarrows B$ chemical reaction on the phase separation in binary mixtures by the generalization (4.55) of the Landau-Ginzburg equation (4.52) in the presence of a chemical reaction. A similar approach has been applied [136] to the isomerization reaction between A and B

$$A \underset{k_2}{\overset{k_1}{\rightleftarrows}} B \tag{4.57}$$

in which C does not participate, and to ternary polymer blends of identical chains lengths with the chemical reaction [137]

$$A + B \underset{k_2}{\overset{k_1}{\rightleftarrows}} C. \tag{4.58}$$

The Flory-Huggins free energy for the ternary mixture has the following form [138],

$$\begin{aligned} f\left(x_A, x_B\right) = {} & x_A \ln x_A + x_B \ln x_B \\ & + \left(1 - x_A - x_B\right) \ln\left(1 - x_A - x_B\right) + n\chi_{AB} x_A x_B \\ & + n\chi_{BC} x_B \left(1 - x_A - x_B\right) + n\chi_{AC} x_A \left(1 - x_A - x_B\right) \end{aligned} \tag{4.59}$$

where x_A, x_B and $x_C = 1 - x_A - x_C$ are the volume fractions of the components, χ is the binary interaction parameter, and n is the number of repeated units in the polymer chain. Two order parameters for these systems are given by

$$\phi = x_A - x_B; \qquad \eta = x_A + x_B - \psi_C \tag{4.60}$$

where ψ_C is the critical composition of the phase diagram for phase separation between C and $A - B$. Expanding (4.59) in the order parameters η and ϕ and their derivatives, one obtains [136]

$$\begin{aligned} F\{\phi, \eta\} = {} & \int dr \left[f_\eta(\eta) + f_\phi(\phi) + f_{int}(\eta, \phi)\right] \\ & + \frac{D_\eta}{2} \int dr \left[\nabla \eta(r)\right]^2 + \frac{D_\phi}{2} \int dr \left[\nabla \phi(r)\right]^2 \end{aligned} \tag{4.61}$$

where D_η and D_ϕ are phenomenological parameters related to the surface energy, and

$$\begin{aligned} f_\eta(\eta) &= -\frac{1}{2}c_1\eta^2 + \frac{1}{4}c_2\eta^4 \\ f_\phi(\phi) &= -\frac{1}{2}c_3\phi^2 + \frac{1}{4}c_4\phi^4 \\ f_{int}(\eta,\phi) &= c_5\eta\phi - \frac{1}{2}c_6\eta\phi^2 + \frac{1}{2}c_7\eta^2\phi^2 \end{aligned} \tag{4.62}$$

where the c_i are functions of the coefficients χ and ψ_C (for details, see [136], [137]). The equation of motion for the order parameters η and ϕ (Landau-Ginzburg equations generalized to a chemically reactive ternary mixture) have a slightly different form for the chemical reactions (4.57) and (4.58). In the former case,

$$\begin{aligned} \frac{\partial\eta}{\partial t} &= M_\eta\nabla^2\frac{\delta F\{\phi,\eta\}}{\delta\eta}, \\ \frac{\partial\phi}{\partial t} &= M_\phi\nabla^2\frac{\delta F\{\phi,\eta\}}{\delta\phi} - (k_1+k_2)\phi \\ &\quad - (k_1-k_2)\eta - (k_1-k_2)\psi_C \end{aligned} \tag{4.63}$$

whereas in the latter case,

$$\begin{aligned} \frac{\partial\eta}{\partial t} &= M_\eta\nabla^2\tfrac{\delta F\{\phi,\eta\}}{\delta\eta} + 2k_2\psi_C - 2k_1x_Ax_B, \\ \frac{\partial\phi}{\partial t} &= M_\phi\nabla^2\tfrac{\delta F\{\phi,\eta\}}{\delta\phi}, \end{aligned} \tag{4.64}$$

where M_ϕ and M_η are the mobility coefficients. For simplicity, the following forms of the functions $f_\eta(\eta)$ and $f_\phi(\phi)$ have been used in numerical simulations,

$$\frac{df_\eta}{d\eta} = A_\eta\tanh\eta - \eta; \qquad \frac{df_\phi}{d\phi} = A_\phi\tanh\phi - \phi \tag{4.65}$$

where A_η and A_ϕ are phenomenological parameters which are inversely proportional to temperature.

Numerical simulations show [136] that for the isomerization reaction (4.57), a difference between phase separation in a ternary mixture and in a binary mixture, related to the availability of the third inert component, leads to the existence of a multiple-phase transition. Otherwise, the dynamics of phase separation in binary and ternary mixtures is quite similar.

For reaction (4.58), the separation dynamics depends on the competition between the chemical reaction and phase separation [137]. In the case of a relatively large reaction rate constant, the effect of a chemical reaction is dominant, phase separation between C and $A - B$ is inhibited, and phase separation occurs between A and B. As the phase separation of A and B proceeds, C decomposes into A and B rapidly, resulting in the phase pattern of small amount of C distributed along the interface between A-rich and B-rich domains. As the reaction rate gradually decreases, the effect of the chemical reaction diminishes. In the later stages of phase separation, the concentration of C increases as the rate constant of the chemical reaction decreases, and the C-rich phase changes from dispersed droplets to a continuous thin phase. In the next section, we consider the dissociation reaction which is a special type of chemical reaction in a ternary mixture.

4.4 Thermodynamics of a three-component plasma with a dissociative chemical reaction

Order-disorder phase transitions and critical phenomena in electrically neutral systems are due to the competition between repulsive and attractive forces. When charged particles appear due to the dissociation-recombination reaction, they give rise to new attractive forces (the Coulomb interaction between ions, which is, on average, an attractive interaction) and repulsive forces (say, the hard-core interaction at short distances). These new competing forces result in an additional first-order phase transition in a three-component plasma with a critical point. The coexisting phases for a new phase transition resulting from the chemical reaction are determined by Eqs. (4.12) and (4.16). It has been proposed [85], [114] that such arguments can be used to explain some experiments in metallic vapors at high pressures and in metal-ammonia solutions.

Metal-ammonia solutions are a type of system in which a small increase of the concentration of metal results in a huge increase of electroconductivity, manifesting a transition from nonmetallic to metallic conductivity. Metallic vapors are another example of the latter behavior. The continuous decrease in the density of metals with heating above the critical temperature causes the transition into the nonmetallic state and a significant decrease of the electric conductivity.

The most general explanation of the nature of this transition was given by Mott [139], who proposed the following elegant physical argument. Mott argued that changing the number density n of charges interacting by a screened Coulomb interaction eliminates the bound states when $a_B n^{-1/3} \approx 0.25$, and hence the system will become metallic. Here, a_B is the Bohr radius in the medium. Despite the generality of the Mott criterion, not all metal-nonmetal transitions can be explained in this way. Sometimes the transition takes place at a density lower than that predicted by the Mott criterion [140]; in other systems, there is more than one transition. Let us consider two examples in more detail.

The liquid-gas critical point of mercury occurs at the density $\rho_C = 5.77$ g/cm^3 ($p_C = 1670$ Bar, $T_C = 1750$ K). Afterwards, a metal-nonmetal transition takes place at $\rho \approx 9$ g/cm^3. It is generally accepted (see, however, [141]) that the latter transition is explained by the Mott criterion. However, a dielectric anomaly has been found [142] at the density $\rho \approx 3$ g/cm^3. This was interpreted as a transition from a weakly ionized mercury plasma to a new inhomogeneous phase of charged droplets. This transition, probably of first order, occurs at densities much lower than those given by Mott, and therefore requires an explanation.

The second example is even more conclusive. The heat capacity of sodium-ammonia solution has been measured [143] for three different molar fractions of metal ($X = 0.045$, 0.0462, and 0.0631). Starting at low temperatures, the heat capacity shows a jump across the liquid-gas separation curve of this mixture. Upon further heating, two additional jumps in the heat capacity occur at two distinct temperatures above the critical temperature. The changes in slope of the electromotive force take place at the same temperatures [144] providing additional evidence for a first-order phase transition above the liquid-gas critical point. Experimental observation of this new phase transition is very difficult because of the very small density differences between the two phases and the existence of slow diffusion processes. The former requires extreme homogeneity of the temperature, and the latter results in very long relaxation times. We believe that these and other experimental problems prevent the observation of this new thermodynamic phase transition for other systems, while its existence seems obvious for a broad class of systems. In addition to metal-ammonia solutions, one can mention molten salts, weak electrolytes, solutions of metals in their salts, electron-hole plasmas in optically excited semiconductors, etc.

A ternary mixture with an ionization reaction is another example of phase separation in reactive systems [114]. Let us consider a system composed of neutral particles which are partially dissociated into positive and negative charges. Assume that the Gibbs free energy has the following form,

$$G = N_A \mu_{A,0} + N_I \mu_{I,0} + N_e \mu_{e,0} + N_A \kappa_B T \ln x_A + 2N_I \kappa_B T \ln x_I + N_I I + \mu_0 N_I x_I^{\alpha} \quad (4.66)$$

where N_A is the number of neutral particles, $N_I\,(N_e)$ is the number of ions (electrons), I is the ionization potential, and $x_A \equiv N_A/(N_A + 2N_I) = 1 - 2x$. Electroneutrality requires that $N_I = N_e$.

The first three terms in Eq. (4.66) describe the ideal gases of the particles. The next two terms are the entropy, the sixth term describes the ionization process, and the last term represents the interaction between the charged particles. We will later make the Debye-Hückel approximation,

$$\alpha = \frac{1}{2}; \qquad \mu_0 = -\frac{e^3}{3\kappa_B T}\left(\frac{p}{2\varepsilon^3 \pi^2}\right)^{1/2}. \quad (4.67)$$

One can formulate the Debye-Hückel approximation using either the Gibbs or the Helmholtz free energy. If one switches from one free energy to the other by using the ideal gas equation of state in the correction term, the Debye-Hückel approximation is slightly different. We here use the Gibbs free energy, and our equation (4.67) follows from Eqs. (94.1) and (75.14) of Landau and Lifshitz [4].

The chemical potentials of the neutral and charged particles can be easily found from (4.66),

$$\begin{aligned} \mu_A &= \mu_{A,0} + \kappa_B T \ln x_A - \alpha \mu_0 x^{\alpha+1}, \\ \mu_q &\equiv \mu_I + \mu_e = \mu_{I,0} + \mu_{e,0} \\ &\quad + 2\kappa_B T \ln x + \mu_0 \left[(\alpha+1)\, x^{\alpha} - 2\alpha x^{\alpha+1}\right] + I. \end{aligned} \quad (4.68)$$

In order to obtain the coexistence curve, one equates the chemical potentials in the two phases, analogously to Eq. (4.5). Solving these two equations for $\alpha\mu_0/\kappa_B T$ yields

$$\frac{\ln(1-2x') - \ln(1-2x'')}{(x')^{\alpha+1} - (x'')^{\alpha+1}} = \frac{2\alpha(\ln x' - \ln x'')}{2\alpha\left[(x')^{\alpha+1} - (x'')^{\alpha+1}\right] - (\alpha+1)\left[(x')^{\alpha} - (x'')^{\alpha}\right]}. \quad (4.69)$$

One can easily verify that Eq. (4.69) gives the following relation between the charge concentrations in the coexisting phases,

$$\left(\frac{x'}{x''}\right)^{\alpha} = \frac{1-2x''}{1-2x'}. \tag{4.70}$$

For the Debye-Hückel interaction (4.67), the coexistence curve has the following form,

$$x' = \left[1 - x'' - (2 - 3x'')^{1/2} (x'')^{1/2}\right]^{1/2},$$

$$-\left[\frac{e^3}{6(\kappa_B T)^2}\right]\left(\frac{p}{2\varepsilon^3\pi^2}\right)^{1/2} = \frac{\ln x' - \ln x''}{(x')^{3/2} - (x'')^{3/2}}. \tag{4.71}$$

If the dissociation reaction $A \rightleftarrows I + e$ takes place, one has to add the restriction stemming from the law of mass action, $\mu_A - \mu_I - \mu_e = 0$. Using (4.67) and (4.68), one can write the latter equation in the following form

$$\mu_A^0 - \mu_I^0 - \mu_e^0 - I + \kappa_B T \ln\frac{1-2x}{x^2} + \frac{\mu_0}{2}\left(3x^{1/2} - x^{3/2}\right) = 0. \tag{4.72}$$

Neglecting the difference between the mass of atom and ion, we put $\mu_A^0 = \mu_I^0$. Furthermore, using the well-known expression for an ideal gas of electrons, one obtains from (4.72),

$$\ln\left[\frac{p}{\kappa_B T}\left(\frac{2\pi\hbar^2}{m_e \kappa_B T}\right)^{3/2}\right] + \frac{I}{\kappa_B T} - \ln\left(\frac{1-2x}{x^2}\right)$$

$$-\left[e^3/6(\kappa_B T)^2\right]\left(\frac{p}{2\varepsilon^3\pi^2}\right)^{1/2}\left(3x^{1/2} - x^{3/2}\right) = 0. \tag{4.73}$$

Equation (4.73) gives the required restriction on the parameters of the coexisting phases resulting from the chemical reaction. One has to find the simultaneous solutions of Eqs. (4.71) and (4.73). Since one cannot solve these equations analytically, one adopts the following approximate procedure: for given I and ε, one starts from some x', which gives x'' and two equations for T and p (for details, see [114]).

Chapter 5

Comments on the Geometry of the Phase Diagram of a Reaction Mixture

5.1 Solubility in supercritical fluids

The properties of fluids slightly above their critical points ("supercritical fluids") came to the attention of researchers more than a century ago, because of their many technological applications. These include supercritical extraction [145], supercritical chromatography [146] , and the strong pressure (or temperature) effect on the reaction rate [26]. We concentrate here on the phenomenon of enhanced solubility of solids in supercritical fluids ("supercritical extraction") [70]. The technological advantages of this phenomenon have aroused great interest in the chemical engineering community. Indeed, this interest is reflected in the existence of a journal (Journal of Supercritical Fluids), books ([147], [148], among others), and hundreds of scientific and technological articles.

Figure 5.1 displays a typical example of the phenomenon. Depicted is the mole fraction of naphthalene dissolved in ethylene as a function of the pressure for various fixed supercritical temperatures [149]. (The critical temperature and pressure of the solvent are 11°C and 51.2 atm.) The same effect was found for many other substances. Figure 5.1 shows that for every isotherm, there is a region of pressure where small changes in pressure result in a huge increase in solubility. Moreover, the slope of this increase becomes larger as T approaches T_C.

The thermodynamic analysis of this phenomenon (similar to that performed in Sec. 2.5) must be appropriate to the experimental setup which allows a solid to equilibrate with a supercritical fluid at a given temperature and pressure. Denote the fluid component (solvent) by subscript 1, and the solid component (solute) by subscript 2. Denoting the solid phase by superscript s and the fluid phase by f, we have $X_2^s = 1$, where X_2 is

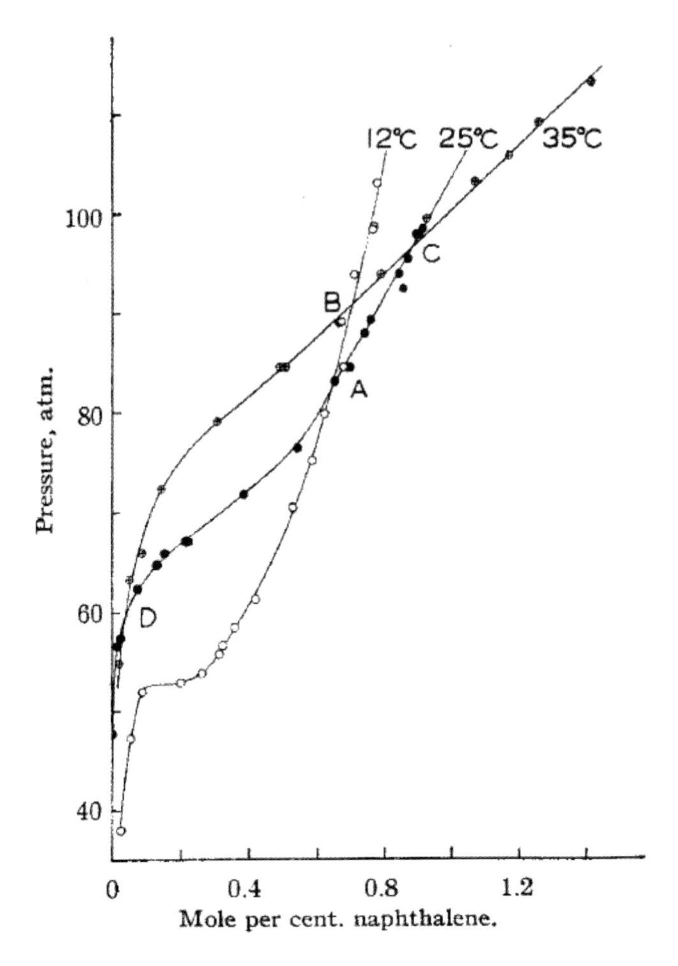

Fig. 5.1 Pressure-naphthalene composition phase diagram of the system ethylene-naphthalene for different temperatures. Reproduced from Ref. [149] with permission, copyright (1948), American Chemical Society.

the mole fraction of the solid component. In the fluid phase, we define $X_2^f \equiv X$. The chemical potential of the solid component is denoted as μ_2 and, accordingly, $\mu_2^s = \mu_2^f$ for the equilibrium state. We choose T, p, X as independent variables. Thus, $\mu_2^f = \mu_2^f(T, p, X)$. For all equilibrium states,

$$\Delta\mu(T, p, X) = \mu_2^s(T, p) - \mu_2^f(T, p, X) = 0 \tag{5.1}$$

using the fact that $X_2^s = 1$ in the solid phase.

Consider isothermal changes of pressure along the equilibrium line,

$$d\left[\Delta\mu\left(T,p,X\right)\right] = \left(\frac{\partial\mu_2^s}{\partial p}\right)_T dp$$

$$-\left(\frac{\partial\mu_2^f}{\partial p}\right)_{T,X} dp - \left(\frac{\partial\mu_2^f}{\partial X}\right)_{T,p} dX = 0 \tag{5.2}$$

Using the definition of the molar volume of the solid $v^s = (\partial\mu_2^s/\partial p)_T$ and of the partial volume of the solute in the fluid phase $\overline{v} = \left(\partial\mu_2^f/\partial p\right)_{T,X}$, one can rewrite Eq. (5.2) in the following form

$$\left(\frac{\partial X}{\partial p}\right)_{T,equil.line} = \frac{v^s - \overline{v}}{\left(\partial\mu_2^f/\partial X\right)_{T,p}}. \tag{5.3}$$

This result is completely general, and will form the basis of the subsequent analysis. One can already see from (5.3) that near the critical points, the denominator goes to zero, leading to the strong divergence of the left-hand side of this equation, which corresponds to the graphs shown in Fig. 5.1.

It is instructive to examine Eq. (5.3) far from criticality. Using the well-known form of the chemical potential of dilute solutions,

$$\mu_2^f = \mu_{2,0}^f\left(T,p\right) + RT\ln X, \tag{5.4}$$

one obtains from (5.3),

$$\left(\frac{\partial\ln X}{\partial p}\right)_{T,equil.line} = \frac{v^s - \overline{v}}{RT}. \tag{5.5}$$

Far from criticality, the fluid mixture is ideal, and the partial volume of the solute $\overline{v}$ is much larger than its molar volume in the solid phase, $\overline{v} >> v^s$. Therefore, $(\partial\ln X, \partial p)_{T,equil.line} < 0$, i.e., X decreases as p increases. In addition,

$$\left(\frac{\partial^2\ln X}{\partial p^2}\right)_{T,equil.line} = \frac{1}{RT}\left[\left(\frac{\partial v^s}{\partial p}\right)_T - \left(\frac{\partial\overline{v}}{\partial p}\right)_{T,X}\right] \tag{5.6}$$

because $\overline{v}$ does not depend on X in this region of pressure. Therefore, the derivatives along the equilibrium line coincide with the derivatives at a point. Since $|(\partial v^s/\partial p)_T| << \left|(\partial\overline{v}/\partial p)_{T,X}\right|$ (i.e., the solid compressibility is almost zero), the equilibrium isothermal curve $X\left(p\right)$ has positive curvature for low pressures. Since $(\partial\overline{v}/\partial p)_{T,X}$ is negative but much larger in absolute value than $(\partial v^s/\partial p)_T$, $\overline{v}$ decreases with increasing pressure and a minimum in the curve $X(p)$ occures when $\overline{v} = v^s$. From this point onward, $X(p)$ is an increasing function unless it reaches the critical region.

Let us now return to the problem of solubility in the critical region. One can immediately see that the chemical potential of a solute in dilute solutions cannot have the usual form (5.4) near the critical points, because one cannot satisfy Eq. (1.2) using the potential of the form (5.4). In this sense [150], the critical dilute solution is not "dilute"! In deriving Eq. (5.4), it was assumed that one can neglect the interaction between solute particles because of the initial assumption that the solution is very weak. However, this is incorrect near the critical point because long-range correlations (1.5) cause the effective "interaction" between the solute particles to be important even for dilute solutions.

Let us obtain the correct expression for the chemical potential μ_s of a solute in a binary mixture (calculations for many-component mixtures are quite analogous). Differentiating the Gibbs-Duhem relation [3] $v = v_1 X + v_2 (1 - X)$ for the partial volumes v_1 and v_2, with respect to X, and eliminating v_2, yields

$$v_1 = v + (1 - X) \left(\frac{\partial v}{\partial X} \right)_{T,p}. \tag{5.7}$$

However, the partial volume is related [3] to the chemical potential $v_1 = (\partial \mu_s / \partial p)_{T,x}$. Integrating the latter formula, and using (5.4) for the ideal gas as a reference system, gives

$$\mu_s (T, p, X) = RT \ln X + \int_0^p (v_1 - v_{id})\, dp. \tag{5.8}$$

Differentiating (5.8) with respect to X yields

$$\left(\frac{\partial \mu_s}{\partial X} \right)_{T,p} = \frac{RT}{X} + \int_0^p \frac{\partial v_1}{\partial X} dp$$

$$= \frac{RT}{X} + (1 - X) \int_0^X \left(\frac{\partial^2 v}{\partial X^2} \right) \left(\frac{\partial p}{\partial X} \right) dX. \tag{5.9}$$

In the last equality in (5.9), we used the derivative of Eq. (5.7) and $dp = (\partial p / \partial X)_{T,v}\, dX$, which follows from the equation of state $p = p(T, v, X)$.

Integration by parts and using simple thermodynamic transformations enables one to rewrite Eq. (5.9) in the final form,

$$\left(\frac{\partial \mu_s}{\partial x} \right)_{T,p} = \frac{RT}{X}$$

$$+ (1 - X) \frac{(\partial p / \partial X)^2_{T,v}}{(\partial p / \partial v)_{T,X}} - (1 - X) \int_0^v \left(\frac{\partial^2 p}{\partial X^2} \right)_{T,v} dv. \tag{5.10}$$

In contrast to (5.4), Eq. (5.10) describes the critical line of a binary mixture, if one assumes that $(\partial p/\partial v)_{T,X}$ and $\left(\partial^2 p/\partial v^2\right)_{T,X}$ are proportional to $-X$. The former condition leads to the cancellation of the singularity RT/X in (5.10), while the latter allows one to satisfy the condition $\left(\partial^2 \mu/\partial X^2\right)_{T,p} = 0$.

After this general comment, let us consider Eq. (5.3), and its form in the critical region. We will show that $(\partial X/\partial p)_{T,equil.line}$ diverges at two different critical points in the phase diagram, called the upper and lower critical end point (UCEP and LCEP), leading to a marked dependence of X on p in their respective critical regions.

We first rewrite Eq. (5.3) in the following form,

$$\left(\frac{\partial X}{\partial p}\right)_{T,equil.line} = \frac{v^s - \left[v - (1-X)\left(\frac{\partial p}{\partial X}\right)_{T,v}\left(\frac{\partial p}{\partial v}\right)_{T,X}^{-1}\right]}{\left(\frac{\partial \mu_2^f}{\partial X}\right)_{T,p}} \tag{5.11}$$

where v is the molar volume, and use has been made of $\overline{v} = v + (1-X)(\partial v/\partial X)_{T,p}$ and $(\partial v/\partial X)_{T,p} = -(\partial p/\partial X)_{T,v}(\partial p/\partial v)_{T,X}^{-1}$. Equation (5.11) contains two "dangerous" derivatives. The derivative $(\partial p/\partial v)_T$ vanishes at the critical point of the pure fluid. On the other hand, the derivative $\left(\partial \mu_2^f/\partial X\right)_{T,p}$ vanishes on the liquid-gas critical line of the binary mixture, which terminates at the pure-fluid critical point. However, the experimental setup considered here is special, and destroys the relevance of the critical line. Due to the contact between the fluid phase and the solid phase, there is a constraint (5.1) on the chemical potential of the solute in the fluid phase. This constraint decreases the number of degrees of freedom of the system, and the system has only isolated critical points rather than a critical line. These critical points can be found from a solution of Eqs. (5.1) and (1.2) which define the critical line of a binary mixture. For typical phase diagrams, these two points are UCEP and LCEP [70]. At these two points, and nowhere else, the denominator of Eq. (5.11) vanishes.

It was found experimentally that the temperature T_C of LCEP is located close to the critical temperature $T_{C,0}$ of the pure, more volatile component. For example, $(T_C - T_{C,0})/T_{C,0} < 10^{-2}$ for all the solution of different compounds in ethylene [151]. Clearly, when $(T_C - T_{C,0})/T_{C,0}$ is small, we also expect X and $(p_C - p_{C,0})/p_{C,0}$ to be small. In such cases, we are faced with a critical point of a mixture that is very dilute and, therefore, is close to the critical point of the solvent, where $(\partial p/\partial v)_T = 0$. Comparing with Eq. (5.11), we conclude that there are two reasons for the divergence of $(\partial X/\partial p)$ near the LCEP (both $(\partial p/\partial v)$ and $(\partial \mu/\partial X)$ tend to zero), but only one reason for the divergence near the UCEP, where only $(\partial \mu/\partial X)$ goes to zero.

A qualitative analysis has been performed separately for LCEP and UCEP [70] under the assumption of analyticity at the critical points. Another approach to supercritical extraction near UCEP and LCEP is based on a scaling theory of the critical phenomena [152]. The slope of the solubility curve (5.3) is defined by the derivative $\left(\partial \mu_2^f/\partial X\right)_{T,p}$,

$$\left(\frac{\partial \mu_2^f}{\partial X}\right)_{T,p} = a\pi + b\tau^{\gamma} + cX^{\delta-1} \tag{5.12}$$

where $\tau \equiv (T - T_C)/T_C$, $\pi \equiv (p - p_C)/p_C$, and a, b, c are constants, while γ and δ are the critical indices [152].

The temperature dependence of (5.12) is defined by the thermodynamic path in approaching the critical point. The experiments were performed at varying concentration. Therefore, it is natural to assume the smoothness of the chemical potential of the solvent and the solute as a function of pressure.

Integrating (5.12), one obtains

$$\mu_2^f = a\pi X + b\tau^{\gamma} X + \frac{c}{\delta} X^{\delta} + b_1\tau + c_1\pi \tag{5.13}$$

where b_1 and c_1 are constants.

For the experimental thermodynamic path $\pi = const$, the compatibility between other variables will be of the form $\tau \sim X^{\delta}$. Therefore, one concludes from (5.13) that asymptotically

$$\left(\frac{\partial \mu_2^f}{\partial X}\right)_{T,p} \sim \tau^{(\delta-1)/\delta}. \tag{5.14}$$

In Eq. (5.12), we neglect the second term having asymptotic behavior τ^{γ} compared to the third term proportional to $\tau^{(\delta-1)/\delta}$, since [152] $(\delta - 1)/\delta < \gamma$. In the asymptotic regime, one keeps the term having the lowest power of τ.

If the LCEP is close to the critical point of the pure solvent, another singularity appears in the slope of the solubility, that is due to the "weak" singularity of the partial molar volume

$$\overline{v} \sim \tau^{-\alpha}. \tag{5.15}$$

Equation (5.15) has been verified experimentally [153].

Our analysis shows that for $p = p_C$ and for different isotherms, the slope of solubility as a function on τ on the log-log scale is a straight line independent of the system considered, with the value of this slope depending on the values of critical indices at different distances from the critical point. A straight line with slope 1.05 seems to correspond to the experiment results. However, we need more precise measurements.

From the large number of results related to supercritical fluids, we bring here a partial list of possible applications of chemical reactions in supercritical fluids [154]–[156]. There are certain characteristic features of supercritical fluids which make these applications especially convenient, including their ability to dissolve various components, in particular, solids, the strong temperature and pressure dependence of the solubility, and the ability to use cheap inorganic fluids. Those who drink decaffeinated coffee and beer probably know that considerable amounts of caffeine and hops are currently produced by CO_2 supercritical extraction. Among many applications in the food and pharmaceutical industries, one can mention the extraction of oils, flowers and fragrances. The rapid motion through a nozzle of an expanded supercritical fluid with dissolved solids leads to the formation of very small monodisperse particles of prescribed size and morphology, which is useful in biotechnology and material science [157]. Other applications include enhanced oil recovery with supercritical CO_2 [158], wet-air oxidation of organics in supercritical water [159], coal liquefaction in supercritical toluene [160], catalytic disproportionation of near-critical toluene [161], bitumen extraction from oil shale [162], activated carbon regeneration [163], and supercritical chromatography [164]. Technological applications (metallurgical, geochemical, geological) can be added to this list (see [165] for dozens of references).

The main interest in diffusion-controlled chemical reactions in supercritical fluids is associated with the large reaction rates, which are several order of magnitude higher than in normal fluids due to the small kinematic viscosity. Moreover, these rates can be strongly modified by a change of thermodynamic parameters. Modern technology also uses another types of chemical reactions, such as enzymatic reactions in supercritical CO_2, oxidation of hazardous materials in supercritical water, and catalyzed heterogeneous reactions.

5.2 Azeotropic points in reactive many-component systems

A typical temperature-composition projection of the $T - x' - x''$ equilibrium surface (at constant pressure) is shown in Fig. 5.2. This mixture forms an azeotrope point A where [19] "distillation (or condensation) takes place without change of composition". Like critical states, azeotropic states do not exist for ideal nonreactive mixtures.

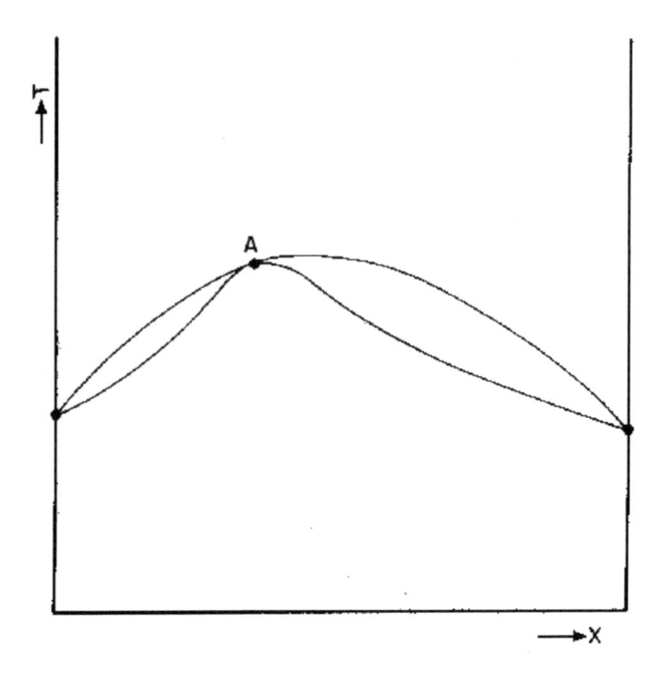

Fig. 5.2 Liquid-gas equilibrium in a binary mixture, showing the azeotrope point A. Reproduced from Ref. [18] with permission, copyright (1990), Springer.

The situation is quite different for reactive mixtures, for which [166] an azeotrope exists even for ideal mixtures, and it is no longer defined by the equality of phase composition. For reactive mixtures, the latter condition for n-component two-phase systems is replaced by

$$\frac{x_1'' - x_1'}{\nu_1 - \nu_T x_1'} = \frac{x_i'' - x_i'}{\nu_i - \nu_T x_i'}; \qquad i = 1, 2, \ldots, n-1 \tag{5.16}$$

where $\nu_T = \sum_{i=1}^{n} \nu_i$.

A simple proof [166] of Eq. (5.16) is based on the conservation laws in a closed systems in which the n-component liquid mixture is vaporized at constant pressure or at constant temperature. The material balance for component i gives

$$\frac{d\left(\upsilon' x_i'\right)}{dt} + \frac{d\left(\upsilon'' x_i''\right)}{dt} = \upsilon_i \frac{d\xi}{dt}; \quad i = 1, 2, \ldots, n-1 \tag{5.17}$$

where υ' and υ'' are the molar volumes of the two phases and ξ is the extent of reaction. The overall material balance

$$\frac{d\upsilon'}{dt} + \frac{d\upsilon''}{dt} = \upsilon_T \frac{d\xi}{dt} \tag{5.18}$$

permits one to rewrite Eq. (5.17) in the following form

$$\upsilon' \frac{dx_i'}{dt} + \upsilon'' \frac{dx_i''}{dt} = \left(\nu_i - \nu_T x_i'\right) \frac{d\xi}{dt} - \left(x_i'' - x_i'\right) \frac{d\upsilon''}{dt}; \quad i = 1, 2, \ldots, n-1. \tag{5.19}$$

During the azeotrope transformation, the composition of each phase remains constant, i.e., $(dx_i'/dt) = (dx_i''/dt) = 0$, and (5.19) reduces to

$$\frac{x_i'' - x_i'}{\nu_i - \nu_T x_i'} = \frac{d\xi}{dt} \left(\frac{d\upsilon''}{dt}\right)^{-1}; \qquad i = 1, 2, \ldots, n-1 \tag{5.20}$$

The right-hand side of (5.20) does not depend on i, which proves Eq. (5.16).

5.3 Melting point of reactive binary mixtures

Consider the phase diagram near the melting point for a two-phase system consisting of a solid component AB which dissociates completely on melting and the liquid of the same composition. The $T - x_B$ projection of the phase diagram is shown in Fig. 5.3, where T^+ is the melting point, the vertical line at $x_B = 1/2$ describes a solid phase, and the dashed lines relate to liquid phases with concentrations larger and smaller than 1/2. The existence of two branches of the solubility curve can be explained by the lowering of the melting point by the addition or removal of one component. Therefore, the solubility lines meet at $T = T^+$.

Let us now consider a chemical reaction (dissociation) in the liquid phase:

$$AB \rightleftarrows A + B. \tag{5.21}$$

The solubility curve will then be rounded at $T = T^+$, as shown in Fig. 5.3 by the solid curves. The solubility curve becomes more rounded with an increase of the degree of dissociation. This may be explained as follows. Let

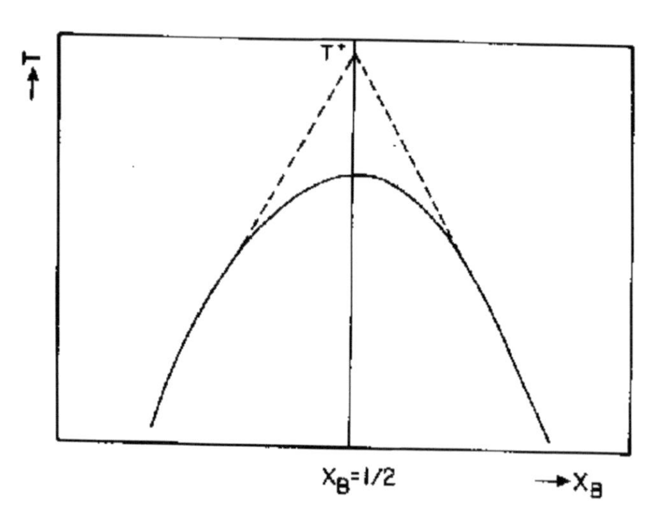

Fig. 5.3 Liquid-solid solubility curves for compound AB with (solid curve) and without (dashed curve) chemical reaction. Reproduced from Ref. [18] with permission, copyright (1990), Springer.

the liquid solution has the total compositions x_A, x_B of substances A and B and detailed compositions y_A, y_B, y_{AB} of A, B and AB ($y_A+y_B+y_{AB}=1$). The chemical potential of AB in the solution must be equal to that in the solid, which plays the role of the chemical potential "bath" for the solution. The chemical potential of the solid depends only on temperature and, therefore, one may assume that for small degrees of dissociation, y_{AB} is a function of T.

Let us now introduce the small parameter ξ to describe the left part of the melting curves ($x_B < 1/2$):

$$\xi = 2\left(\frac{1}{2} - x_B\right). \tag{5.22}$$

The x, y variables are connected by

$$x_B = \frac{y_B + y_{AB}}{1 + y_{AB}}. \tag{5.23}$$

Using Eqs. (5.22)–(5.23), and the relation $y_{AB} = f(T)$, one obtains

$$\begin{aligned} y_A &= \frac{1 - f(T)}{2} + \frac{\xi}{2}\left[1 + f(T)\right]; \\ y_B &= \frac{1 - f(T)}{2} - \frac{\xi}{2}\left[1 + f(T)\right]. \end{aligned} \tag{5.24}$$

If dissociation is slight, than the solution is ideal. Therefore, one can write the law of mass action for the reaction (5.21) as follows,

$$y_A y_B = K(T). \tag{5.25}$$

Inserting (5.24) into (5.25), one can rewrite the latter in the form

$$\xi^2 = \frac{[1 - f(T)]^2}{[1 + f(T)]^2} - \frac{4K(T)}{[1 + f(T)]^2} \equiv Q(T)^2 - R(T)^2. \tag{5.26}$$

At $T = T^+$, $\xi = 0$, i.e., $Q(T^+) = R(T^+)$. Expanding Eq. (5.26) near $T = T^+$ yields

$$\xi^2 = 2Q(T^+)\left[\left(\frac{dR}{dT}\right) - \left(\frac{dQ}{dT}\right)\right]_{T=T^+}(T^+ - T) + \cdots. \tag{5.27}$$

Equation (5.27) describes a melting line in the presence of a chemical reaction (solid curve in Fig. 5.3).

We now turn to the case of a nonreactive mixture. Then, $K(T) = 0$, and the first term in the series expansion of Eq. (5.26) near $T = T^+$ is of second order,

$$\xi^2 = \left(\frac{dQ}{dT}\right)^2_{T=T^+}(T^+ - T)^2. \tag{5.28}$$

In this case the melting lines (the dashed curves in Fig. 5.3) meet at a point.

The reason for the difference between Eqs. (5.27) and (5.28) was explained back in 1892 [167]. If there were any appreciable dissociation of AB into its constituents, then, in contrast to the non-dissociated solution, the small addition of either component will not change the equilibrium temperature. Therefore, the melting point is a true maximum of a solid curve [Eq. (5.27)] and not a kink [Eq. (5.28)].

Similar results are obtained from more general considerations, including the more complicated compounds $A_m B_n$ [168] and the strictly regular solution model [169] (instead of an ideal model).

Further development of these ideas has been carried out by Krichevskii et al. [170], [171]. They considered the influence of a third component on the thermodynamics of the three-phase equilibrium among a solid phase (compound AB_m, a solvent B with a dissolved substance A), a liquid phase (solution saturated with this component and extremely dilute with respect to component C), and a gas phase (vapor of the pure solvent B with the components A and C assumed to be nonvolatile).

For a non-dissociated system, the change in fugacity of the solvent B (which, for the low saturated vapor pressure, coincides with the partial pressure) due to the addition of a small amount of C at constant pressure and temperature, is given by [172]

$$\ln \frac{p_{B,(ABC)}}{p_{B,(AB)}} = -\frac{n_C}{n_B} \tag{5.29}$$

where $p_{B,(AB)}$ and $p_{B,(ABC)}$ are the partial pressures of the solvent in binary and ternary solutions at the same temperature, and n is the number of moles of the components.

Let us now allow the chemical reaction (dissociation) in the solution,

$$AB_m \rightleftarrows A + mB. \tag{5.30}$$

The chemical potential of the compound AB_m in solution, which coexists with a solid, is fixed and depends only on the temperature. The fugacity γ of the nonvolatile component A can be replaced by the activity a. Therefore, the law of mass action for the reaction (5.30) can be written as

$$a_A \gamma_B^m = K \tag{5.31}$$

or, in differential form,

$$d\,(\ln a_A) + m\; d \ln\,(\gamma_B) = 0. \tag{5.32}$$

Combining this equation with the Gibbs-Duhem equation [3]

$$n_A\; d\,(\ln a_A) + n_B\; d\,(\ln a_B) + n_C\; d\,(\ln a_C) = 0 \tag{5.33}$$

one obtains

$$d \ln\,(\gamma_B) = -\frac{n_C d\,(\ln a_C)}{n_B - m n_A} \tag{5.34}$$

If the solution is very dilute with respect to C, one can replace the activity a_C by the numbers of moles n_C.

Integrating Eq. (5.34) yields

$$\ln \frac{\gamma_{B,(ABC)}}{\gamma_{B,(AB)}} = \ln \frac{p_{B,(ABC)}}{p_{B,(AB)}} = -\frac{n_C}{n_{B,0} - m n_{A,0}} \tag{5.35}$$

For the chemical reaction of the form (5.30), the difference $n_B - m n_A$ remains constant. Consequently, we replace n_B and n_A by the total number of moles of the components $n_{B,0}$ and $n_{A,0}$ in both liquid and solid phases.

At $m = 0$, Eq. (5.35) reduces to Eq. (5.29). At the melting point in Fig. 5.3 (drawn for $m = 1$), where two branches of the solubility curves meet, the compositions of the solid and liquid phases are equal, and the denominator of Eq. (5.35) reduces to zero. Addition of component C to solution must decrease the value of $p_{B,(ABC)}$ on the left branch ($n_B > mn_A$) and increase it on the right branch ($n_B < mn_A$).

The immediate vicinity of the melting point is of special interest. On approaching the melting point along the left branch, the right-hand side of Eq. (5.35) tends to $-\infty$, whereas approaching along the right branch, it tends to $+\infty$. Correspondingly, $p_{B,(ABC)}$ must decrease to zero and increase to infinity, respectively. Both these values (zero and infinity) are physically impossible. In fact, $p_{B,(ABC)}$ on the left branch may not be less than the dissociation pressure of the compound in solution, and on the right branch, $p_{B,(ABC)}$ cannot exceed the saturated vapor pressure of the pure solvent.

5.4 Double critical point

The double critical (or hypercritical or re-entrant) point (DCP) results from merging the phase-separation region with the upper (UCST) and lower (LCST) critical solution points. A binary mixture possessing a LCST separates into two phases as the temperature is raised above the LCPT, and is miscible in all proportions below LCST. In contrast, the mixture is miscible in all proportions above UCST. By merging these two points into one "double critical point", the coexistence curve has a closed-loop form, separating the closed ordered region from the single-phase disordered region. Therefore, the ordered two-phase state exists in the limited temperature interval between LCST and UCST [173] shown in Fig. 5.4 together with the DCP at a certain concentration x_0.

Such phase diagrams have been found in binary mixtures of glycerol-guaiacol [174] and of water with nicotine [175], lutidine [176], and other organic amines and hydroxyethers. The main interest in the phase diagrams of the type shown in Fig. 5.4 comes from crystals of the technologically important Rochelle salt group [177]. In multicomponent systems (with three and more components), there are more than two coexistence phases, creating third-order and higher-order critical points. In some many-component mixtures, such as propanol-water-NaCl, there is a line of double critical points which divides the phase separation surface into two parts with UCST and LCST, respectively.

We present here two possible descriptions of the double critical point, based on the mean-field model and on the phenomenological approach. Consider [117] a ternary mixture with mole fractions x_A, x_B, and x_C, which undergoes a chemical reaction $A + B \rightleftarrows C$. Assume that A and B form a

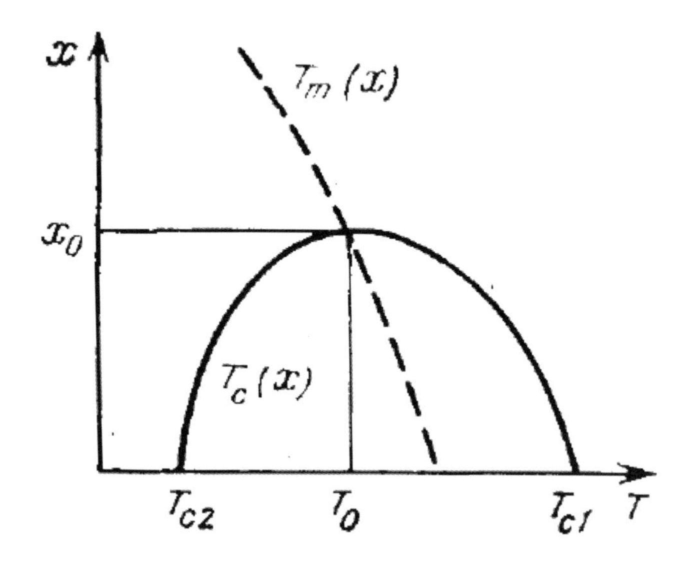

Fig. 5.4 Lines of the upper (UCST) and lower (LCST) critical solution points which merge at concentration x_0. Reproduced from Ref. [173] with permission, copyright (1986), Turpion Publications.

highly directional and energetically favorable bond in becoming C with a positive interaction of mixing $W > 0$, and all other interactions included in the standard chemical potentials $\mu_i^0(p, T)$ of each species. In the absence of a chemical reaction, the coexistence surface for this mixture contains [178] a line of UCST's that decreases to lower temperatures as the concentration of C is increased. However, as will be shown, in the presence of a chemical reaction, along with UCST, the LCST will appear leading to the closed-loop coexistence curve. As it was described in Sec. 4.2.2, in the mean-field approximation, the chemical potentials of the species have the following form

$$\begin{aligned}
\mu_A &= \mu_A^0(p, T) + \ln x_A + (W/k_B T)\, x_B (1 - x_A), \\
\mu_B &= \mu_B^0(p, T) + \ln x_B + (W/k_B T)\, x_A (1 - x_B), \\
\mu_C &= \mu_B^0(p, T) + \ln x_C - (W/k_B T)\, x_A x_B.
\end{aligned} \tag{5.36}$$

The nonlinear terms in (5.36) are symmetric with respect to interchange of A and B, leading to the coexistence phases $x'_A = x''_B$, $x''_A = x'_B$ and $x'_C = x''_C$. Therefore, for the phase equilibrium surface, one gets

$$\frac{W}{k_B T} = \frac{1}{x_A - x_B} \ln \frac{x_A}{x_B}. \tag{5.37}$$

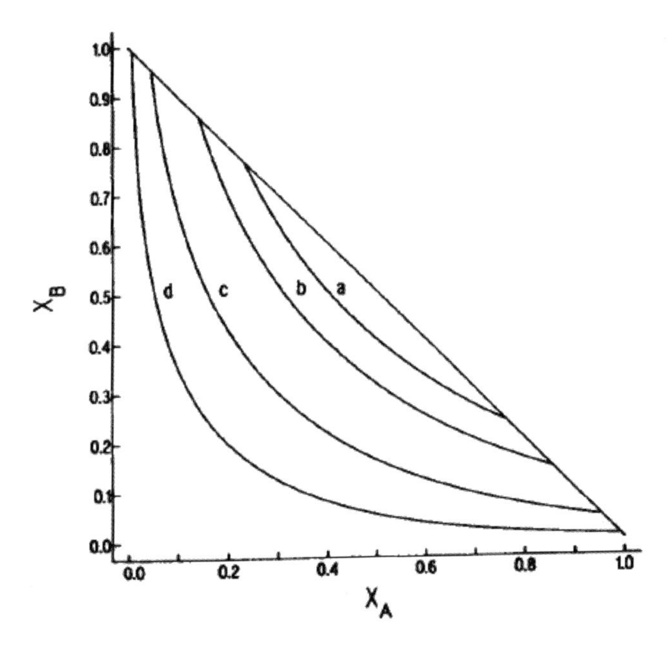

Fig. 5.5 Isotherms of the phase equilibrium surface in a mean-field model. Curves a, b, c, and d correspond to $k_BT/W = 0.45$, 0.4, 0.3 and 0.2, respectively. Reproduced from Ref. [117], copyright (1989), American Institute of Physics.

The set of isotherms in the $x_A - x_B$ plane is shown in Fig. 5.5.

With a decrease in temperature, the phase equilibrium surface approaches the x_A and x_B axes, whereas with increasing temperature, one approaches the critical point defined by $T_{cr} = W/2k_B$, $x_{A,cr} = x_{B,cr} = 1/2$, $x_{C,cr} = 0$. Let us now consider the constraint imposed by the chemical equilibrium,

$$\mu_A + \mu_B - \mu_C = 0 \tag{5.38}$$

Equations (5.38) and (5.36) define the equilibrium expression for x_C,

$$x_C = x_A\, x_B\, \exp\left[-\frac{\Delta H^0}{k_BT} + \frac{\Delta S^0}{k_B}\right] \exp\left[\frac{W}{k_BT}\,(x_A + x_B - x_Ax_B)\right] \tag{5.39}$$

where ΔH^0 and ΔS^0 are the enthalpy and entropy of the chemical reaction which are assumed to be constant. The solutions of Eqs. (5.37) and (5.39) define the coexistence surface of the reactive system in the plane $(k_BT/W, x_A)$ for fixed values of $\Delta H^0/W$ and $\Delta S^0/k_B$. These equations have been solved numerically [117] for $\Delta H^0/W = -10$ and several values of $\Delta S^0/k_B$. The resulting phase diagram (Fig. 5.6) shows that $A - B$ repulsion results in phase separation as the temperature is lowered. However, upon lowering the temperature still further, the formation of C becomes favorable, and the phases become miscible again. In this manner, one obtains the closed-loop coexistence curves created by LCST and UCST with their ratio decreasing as $-\Delta S^0/k_B$ decreases.

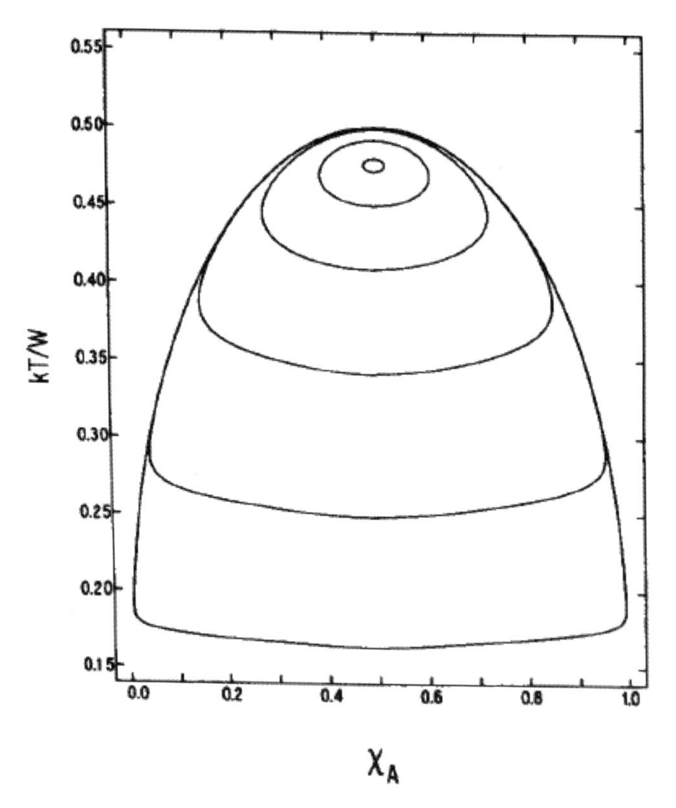

Fig. 5.6 Closed-loop coexistence curves for $\Delta H^0/W = -10$ and $(-\Delta S^0/k_B) = 60$, 40, 30, 26, 24.5, and 24.1. The loops get smaller with decreasing $(-\Delta S^0/k_B)$. Reproduced from Ref. [117], copyright (1989), American Institute of Physics.

As usual, the phenomenological description of the vicinity of the DCP is given by the Landau-Ginzburg expansion for the free energy Φ in the order parameters Ψ and its spatial derivatives:

$$\Phi = \Phi_0 + A(T,x)\,\Psi^2 + B(T,x)\,\Psi^4 + C\,(\nabla\Psi)^2 + \cdots, \qquad (5.40)$$

where the coefficient A changes sign on the line of phase transitions $T_C(x)$, with $A > 0$ for a disordered phase and $A < 0$ for an ordered state. The assumption of analyticity, used in (5.40), is usually applied also to the function $A(T,x)$, namely, $A(T,x) = a(T - T_C)$. It is obvious that the latter assumption is not applicable to the case of DCP, since the non-monotonic behavior of $T_C(x)$ leads to a non-monotonic form of $A(T)$ rather than the linear function [173] (Fig. 5.7).

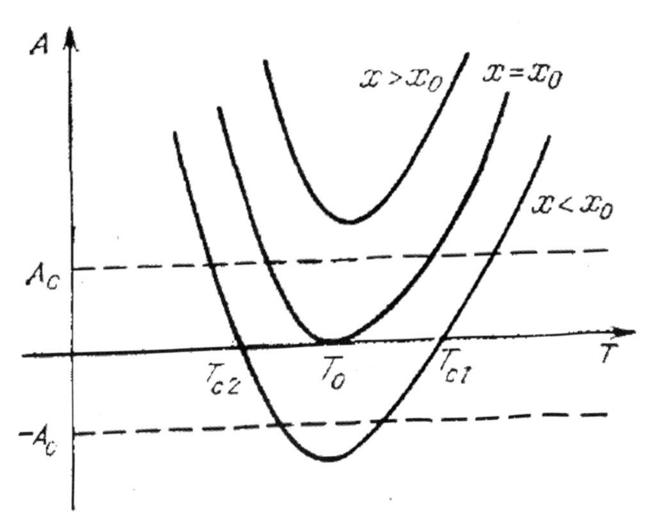

Fig. 5.7 The non-monotonic dependence of the first coefficient in the Landau-Ginzburg expansion as a function of temperature. There are two phase transitions for $x < x_0$, and no phase transitions for $x > x_0$. Reproduced from Ref. [173] with permission, copyright (1986), Turpion Publications.

As one can see from this figure, for $x = x_0$, the curve $A(T)$ is tangent to the temperature axis at the point T_0 , whereas for $x < x_0$, the minimum of $A(T)$ occurs below the horizontal axis, intersecting it at LCST and at UCST. Due to the existence of minima, it is convenient to expand the function $A(T,x)$ around the minimal temperature $T_{\min}$ and not around T_C as is usually done,

$$A(T,x) = A_0(x) + A_2(x)\left|\frac{T - T_{\min}}{T_0}\right|^2 + \cdots . \quad (5.41)$$

The existence of terms in (5.41) quadratic in $|(T - T_{\min})/T_0|$, instead of the usual linear terms, does not change the critical indices, which define the singularities of the compressibility, correlation radius and the quantities related to them. However, the anomalies of the quantities containing the derivatives of the coefficient A with respect to temperature, decrease near the DCP. For example, the jump of the heat capacity $\Delta C = -T\partial^2 \left(A^2/2B\right)/\partial T^2$ at LCST and UCST decreases proportional to the square of the relative distance between these points. Indeed, a decrease of several orders of magnitude in the anomalies of the heat capacity was found in all systems with the DCP. However, there are some discrepancies between the theory and experiment [173] which makes specially interesting the microscopic description of the DCP and the close-loop coexistence curves which have been carried out in the context of lattice-gas [179] and decorated-lattice-gas [96] models.

It should be noted that the anomaly of the kinetic coefficients are also different from the usual critical point. For example, near DCP, the measurements of the shear viscosity of ternary mixtures of 3-methylpyridine, water and heavy water as a function of temperature show [180] that near DCP, the exponent describing the power-law divergence in temperature of the viscosity nearly doubles as the concentration approaches the double-critical-point concentration. There are also experimental data describing light scattering and sound propagation near the DCT [181]. However, we are unaware of any experiments of chemical reactions near DCP, which are of special interest.

Chapter 6

Sound Propagation and Light Scattering in Chemically Reactive Systems

6.1 Ultrasound attenuation in near-critical reactive mixtures

The idea of using the attenuation of sound to study the rate of dissociation of the reaction $N_2O_4 \rightleftarrows 2NO_2$ was proposed in 1920 by Albert Einstein [182]. The total attenuation per wavelength $\alpha\lambda$ near the critical point can be written as

$$\alpha\lambda = (\alpha\lambda)_{cr} + (\alpha\lambda)_{chem}\,, \tag{6.1}$$

neglecting the small background contribution due to viscosity and thermal conductivity. Equation (6.1) contains the assumption of the additive contributions of $(\alpha\lambda)_{cr}$ and $(\alpha\lambda)_{chem}$, which come from the critical fluctuations and chemical reaction, respectively. The assumption of independence of these two contributions is based on the following picture [183]: the reaction takes the system from equilibrium to the steady state, which afterwards will relax to the initial equilibrium state with characteristic relaxation time τ_{chem}. The latter defines the sound attenuation due to a chemical reaction

$$(\alpha\lambda)_{chem} = A_0 \frac{\omega\tau_{chem}}{1 + (\omega\tau_{chem})^2} \tag{6.2}$$

which depends on the nearness to the critical point, since τ_{chem} is proportional to the equilibrium derivative of the extent of reaction ξ with respect to the affinity A, $\tau_{chem} \sim (\partial\xi/\partial A)_{eq}$ [36].

The calculation of the critical contribution to the sound attenuation, expressed by the $(\alpha\lambda)_{cr}$ term in Eq. (2.77), has been performed [184]. This theory is based on an assumption going back to Laplace [185], who corrected Newton calculation showing that the sound velocity in air should be calculated using the adiabatic rather than the isothermal compressibility. In line with this comment, it was assumed [184] that the change of pressure caused by the propagating sound, produces a shift in the critical temperature, but not in the critical concentration. Neglecting the changes in concentration, we are left with two thermodynamic variables p and T, or p and $\tau \equiv T - T_C(p)$, where $T_C(p)$ is the pressure-temperature critical line. Simple thermodynamic analysis shows [184] that the sound velocity u remains constant at the critical point, $u(T = T_C) \equiv u_C$, with small corrections Δu near the critical point,

$$u \equiv u_C + \Delta u = u_C + \frac{g^2 u_C^3}{2T_C C_p} \tag{6.3}$$

where $g = T_C (\partial S/\partial p)_{\lambda\text{-line}} V_C^{-1}$ is a dimensionless constant, and C_p is the specific heat. The above analysis has been performed at non-zero frequency ω for an input pressure signal with time dependence $\exp(-i\omega t)$. Therefore, Eq. (6.3) determines the complex sound velocity $u(\omega)$ as a function of the complex specific heat $C_p(\omega)$. Dynamic scaling theory has been used to determine the frequency dependence of $C_p(\omega)$ from the known temperature dependence of the thermodynamic heat capacity C_p. The function $C_p(\omega)$ can be divided into the noncritical frequency-independent background component C_1 and the critical part $C(\omega)$, where $C_p = C_1 + C(0)$. The low-frequency expansion of $C_p(\omega)$ gives

$$C(\omega) \approx C(0) + \omega C'(0) = C(0)(1 + i\omega\tau) \tag{6.4}$$

where $\tau = -iC_1'(0)/C(0)$ is an effective mean relaxation time. Upon inserting (6.4) into (6.3) and assuming $\omega\tau << 1$, one obtains for the negative imaginary part of the critical velocity Im (u_C) (the sound attenuation per wavelength $\alpha_{cr}\lambda$),

$$\alpha_{cr}\lambda = C(0) \frac{u_C^3}{2T_C} \left(\frac{g}{C_p}\right)^2 \omega\tau. \tag{6.5}$$

The critical part of the specific heat $C(0)$, induced by concentration fluctuations, has the following thermodynamic temperature dependence

$$C(0) \approx t^{-\alpha_0} \tag{6.6}$$

where α_0 is the critical index and $t = \Delta T/T_C = (T - T_C)/T_C$ is the reduced temperature. The relaxation rate of concentration fluctuations of wavenumber k is $k^2 D$, where the critical diffusion coefficient is $D(\xi) = k_B T_C / 6\pi\eta\xi$ with $\xi \approx t^{-\nu}$ and $\eta \approx t^{-z_0}$. Therefore, the characteristic relaxation time for the fluid is

$$\gamma(t) = 2D\xi^{-2} = \gamma_0 (\Delta T)^{\nu z_0} \tag{6.7}$$

where γ_0 is constant. Equation (6.7) enables us to eliminate t in favour of γ as the variable that indicates how close the system is to the critical point,

$$C(\gamma, \omega = 0) = B\left(\frac{\gamma_0}{\gamma}\right)^{\alpha_0/\nu z_0}. \tag{6.8}$$

According to the dynamic-scaling requirement [186], the dependence of Eq. (6.8) on γ disappears when, for $\gamma < \omega$, γ is replaced by $-i\omega$ so that the frequency-dependent specific heat at the critical point is

$$\begin{aligned} C(\gamma = 0, \omega) &= B\left(\frac{\gamma_0}{-i\omega}\right)^{\alpha_0/\nu z_0} \\ &= B \exp\left(\frac{i\pi\alpha_0}{\nu z_0}\right)\left(\frac{\gamma_0}{\omega}\right)^{\alpha_0/\nu z_0}. \end{aligned} \tag{6.9}$$

Replacing $C(0)$ in Eq. (6.5) by the imaginary part of $C(\gamma = 0, \omega)$, and using $\alpha_0 << \nu z_0$, one gets finally

$$\alpha_{cr}\lambda \simeq \frac{\pi^2 \alpha_0 g^2 u_C^2 B}{2T_C C_p^2(T_c)}\left(\frac{\gamma_0}{2\pi}\right)^{\alpha_0/\nu z_0} f^{-\alpha_0/\nu z_0} \tag{6.10}$$

where $f = \omega/2\pi$ is the frequency. Multiplying Eq. (6.10) by $(u_C f)^{-1}$ predicts a linear proportionality between $\alpha_{cr} f^{-2}$ and $f^{-1-\alpha_0/\nu z_0}$. Precisely this dependence had been observed in the measurements [184] of the sound attenuation in 3-methylpentane-nitroethane at various frequencies ranging from 16.5 to 165 MHz. Agreement between theory and experiment was also found for polypropylene-glycol and polyethylene-glycol mixtures [188], iso-butoxyethanol-water system [189], ethanol-dodecane and methanol-cyclohexane mixtures [190], cyclohexane-nitroethane[191], methanol-hexane [192], and triethylamine-water, methanol-cyclohexane [193].

6.2 Hydrodynamic analysis of the dispersion relation for sound waves

Sound waves are oscillating movements of small amplitude, which give rise to small perturbations in the system. Therefore, one can use the linearized hydrodynamic equations for the analysis of the sound velocity and attenuation coefficient in fluids. The linearized system of hydrodynamic equations for reactive systems was obtained in Chapter 2. We will here use the following simplified version of these equations [194]

$$\frac{\partial \rho}{\partial t} = -\rho_0 div\,(v) \tag{6.11}$$

$$\rho_0 \frac{\partial v}{\partial t} = -grad\,(p) + \eta \nabla^2 v + \left(\frac{\eta}{3} + \zeta\right) grad\,(div\,(v)) + \rho_0 L_1 grad\,(A) \tag{6.12}$$

$$\frac{\partial \xi}{\partial t} = -L_0 A - L_1 div\,(v) \tag{6.13}$$

$$A = \left(\frac{\partial A}{\partial \rho}\right)_{\xi,S} \rho + \left(\frac{\partial A}{\partial \xi}\right)_{\rho,S} \xi + \left(\frac{\partial A}{\partial S}\right)_{\xi,\rho} S \tag{6.14}$$

$$p = \left(\frac{\partial p}{\partial \rho}\right)_{\xi,S} \rho + \left(\frac{\partial p}{\partial \xi}\right)_{\rho,S} \xi + \left(\frac{\partial p}{\partial S}\right)_{\xi,\rho} S \tag{6.15}$$

$$T = \left(\frac{\partial T}{\partial \rho}\right)_{\xi,S} \rho + \left(\frac{\partial T}{\partial \xi}\right)_{\rho,S} \xi + \left(\frac{\partial T}{\partial S}\right)_{\xi,\rho} S. \tag{6.16}$$

In contrast to Eqs. (2.61)–(2.64), the independent variables in the above equations are the velocity v, the deviations from equilibrium values of entropy s, density ρ and extent of reaction ξ. Kinetic coefficients are taking into account only in the Navier-Stokes equation (6.12).

To derive the dispersion law, we take the Fourier transform of the hydrodynamic variables

$$\hat{X}\,(\mathrm{K}, \omega) = \int d^3 r \int dt x\,(r, t) \exp\,(i\omega t - i\mathrm{K}r)\ . \tag{6.17}$$

The wave vector K is written $K = k + i\gamma$, where k is the propagation vector and γ is the attenuation coefficient.

After some additional simplifications [194], Eqs. (6.11)–(6.16) yield the following expressions for k and γ

$$k = \frac{\omega}{C}; \qquad \gamma = \frac{\omega^2}{2\rho_0}\frac{G}{C^3} \tag{6.18}$$

where

$$C^2 \equiv -\frac{1}{\rho_0^2}\left[\left(\frac{\partial p}{\partial v}\right)_{\xi,S} + \frac{\left(\frac{\partial p}{\partial \xi}\right)_{v,S}\left(\frac{\partial \xi}{\partial v}\right)_{A,S}}{1+\omega^2\tau^2} + 2\rho_0 L_1 \left(\frac{\partial p}{\partial \xi}\right)_{v,S} \frac{\omega^2\tau^2}{1+\omega^2\tau^2}\right] \tag{6.19}$$

$$G \equiv \left(\frac{\partial p}{\partial \xi}\right)_{v,S}\left[\left(\frac{\partial \xi}{\partial v}\right)_{A,s}\frac{\tau}{\rho_0\left(1+\omega^2\tau^2\right)}\right.$$

$$\left.+\frac{2\tau L_1}{1+\omega^2\tau^2}\right]+\frac{4\eta}{3}+\zeta \tag{6.20}$$

and the relaxation time τ is given by

$$\tau \equiv \left[L_0\left(\frac{\partial A}{\partial \xi}\right)_{v,S}\right]^{-1}. \tag{6.21}$$

Some limit cases of these general equations has been considered [194].

This theory contains visco-reactive cross coefficients L_1, which link the chemical reaction to the momentum flow, which is important for a weak electrolyte solution [194]. Now let us turn to the slightly different hydrodynamic approach, which allows us to consider the vicinity of the critical points without the L_1 terms [195]. Starting from the same hydrodynamic equations obtained in Chapter 2, we replace them by the statistically independent variables as it was done there. However, instead of variables ξ, p, $div\,(v)$ and $\phi = C_{p,x}T/T_0 - \alpha_T p/\rho_0$, we use the set ξ, p, $div\,(v)$ and $S_1 = S - (\partial S/\partial \xi)\,\xi$. The hydrodynamic equations (2.91)–(2.94) may then be rearranged to give

$$\frac{\partial \rho}{\partial t} = -\rho_0 div\,(v) \tag{6.22}$$

$$\rho_0\frac{\partial div\,(v)}{\partial t} = -\nabla^2 p + \left(\frac{4}{3}\eta+\zeta\right)\nabla^2 div\,(v) \tag{6.23}$$

$$\frac{\partial \xi}{\partial t} = D\left[\nabla^2\xi + \frac{k_T}{T_0}\nabla^2 T + \frac{k_p}{p_0}\nabla^2 p\right]$$

$$+\frac{1}{\tau_{T,p}}\left[\xi - \left(\frac{\partial \xi}{\partial T}\right)_{p,A}T - \left(\frac{\partial \xi}{\partial p}\right)_{T,A}p\right] \tag{6.24}$$

$$\frac{\partial S_1}{\partial t} = \frac{\lambda}{\rho_0 T_0}\nabla^2 T + D\frac{k_T}{T_0}\left(\frac{\partial A}{\partial \xi}\right)_{T,p}\left[\nabla^2\xi + \frac{k_p}{p_0}\nabla^2 p\right]$$

$$+D\left(\frac{\partial S}{\partial \xi}\right)_{p,T}\left[\nabla^2\xi + \frac{k_T}{T_0}\nabla^2 T + \frac{k_p}{p_0}\nabla^2 p\right] \tag{6.25}$$

where $\tau_{T,p} = \left[L\left(\partial A/\partial\xi\right)_{T,p}\right]^{-1}$ is the relaxation time, and all other notation has been explained in Chapter 2.

Since only the fluctuations of concentration and entropy become anomalously large near the critical point, we may use the simplified macroscopic equations (6.24)–(6.25) in which the pressure is kept constant. Introducing both normalized and statistically independent spatial Fourier transform variables, one obtains

$$\beta_1(k,t) = \left(\frac{\rho_0}{k_B v C_{p,\xi}}\right)^{1/2} S_1(k,t),$$
$$\beta_2(k,t) = \rho_0\left[\frac{(\partial A/\partial\xi)_{T.p}}{k_B T v}\right]^{1/2} \xi(K,t). \tag{6.26}$$

It is easy to show [195] that

$$\frac{d}{dt}\begin{pmatrix}\beta_1(K,t)\\ \beta_2(K,t)\end{pmatrix} = M\begin{pmatrix}\beta_1(K,t)\\ \beta_2(K,t)\end{pmatrix} \tag{6.27}$$

where the 2×2 matrix M is given by

$$M = \begin{pmatrix} \dfrac{\lambda K^2}{\rho_0 C_{p,\xi}} + \dfrac{\gamma_T}{\tau_{T,p}}, & k_T D\sqrt{\dfrac{\left(\frac{\partial A}{\partial\xi}\right)_{T.p}}{T C_{p,\xi}}}K^2 - \dfrac{\gamma_T^{1/2}}{\tau_{T,p}} \\ k_T D\sqrt{\dfrac{\left(\frac{\partial A}{\partial\xi}\right)_{T.p}}{T C_{p,\xi}}}K^2 - \dfrac{\gamma_T^{1/2}}{\tau_{T,p}}, & DK^2 + \tau_{T,p}^{-1} \end{pmatrix} \tag{6.28}$$

and

$$\gamma_T \equiv \frac{C_{p,A} - C_{p,\xi}}{C_{p,\xi}}. \tag{6.29}$$

The non-diagonal elements of the matrix M are very small and may be neglected if the propagation number K satisfies the inequality

$$K > \left[\frac{C_{p,A} - C_{p,\xi}}{\left(\frac{\lambda}{\rho_0} - DC_{p,\xi}\right)\tau_{T,p}}\right]^{1/2}. \tag{6.30}$$

Numerical estimates indicate that (6.30) is satisfied for $K > 10^4$ cm $^{-1}$. The variables $\beta_1(k,t)$ and $\beta_2(k,t)$ are then effectively decoupled, and

$$\langle\xi(K,t)\,\xi(-K,t)\rangle = \exp\left[-\left(DK^2 + \tau_{T,p}^{-1}\right)t\right]\langle\xi(K)\,\xi(-K)\rangle. \tag{6.31}$$

The mode-coupling method has been used [102] in the analysis of the propagation of sound in a non-reactive binary mixture near the critical point. The general formula for the complex sound attenuation coefficient $\widehat{\alpha}(\omega)$ was obtained in [102],

$$\hat{\alpha}(\omega) \approx \omega^2 \int dK \left(\frac{\partial \ln \langle \xi(K) \xi(-K) \rangle}{\partial T} \right)^2 \times \int_0^\infty dt \exp\left[-2 \left(DK^2 + \tau_{T,p}^{-1} \right) t \right] \exp(i\omega t). \quad (6.32)$$

The arguments that lead to Eq. (6.32) remain unchanged when a chemical reaction occurs. The existence of a chemical reaction results in an additional mechanism for the relaxation of the composition back to its equilibrium value, which results in the appearance of the additional $\tau_{T,p}^{-1}$ term in Eq. (6.31). Then, in a reactive system,

$$\hat{\alpha}(\omega) \approx \frac{\omega^2}{2\left[D(K) K^2 + \tau_{T,p}^{-1}(K) \right] - i\omega} \times \int dK \left(\frac{\partial \ln < \langle \xi(K) \xi(-K) \rangle}{\partial T} \right)^2 \quad (6.33)$$

where the K-dependence of the diffusion coefficient, $D(K)$, and of the relaxation time $\tau_{T,p}^{-1}(K) = L\chi_K^{-1}(K)$ (which reduces to $L(\partial A/\partial \xi)_{T,p}$ for $K = 0$) have been taken into account.

For non-reactive systems, relaxation is associated with diffusion. Just as diffusion processes show critical slowing-down with time scale $\tau_D \sim (Dk^2)^{-1} \sim [(T - T_C)/T_C]^{-3\nu}$, the sound attenuation will show critical behavior over the same time scale. On the other hand, if the chemical reaction occurs on a faster time scale than diffusion and shows critical slowing-down, then the sound attenuation will relax on a time scale $\tau_C \sim \left[(\partial A/\partial \xi)_{T,p} \right]^{-1} \sim [(T - T_C)/T_C]^{-\gamma}$. Because $\gamma < 3\nu$, the chemistry is faster and will dominate the diffusion which can be seen in sound experiments. However, the diffusion coefficient $D(k)$ is wavenumber-dependent and, according to Eq. (3.32), at high frequencies, $D \sim k$. The characteristic diffusion time scale will then be $\tau_D \sim (Dk^2)^{-1} \sim k^{-3}$. This will occur on a faster time scale than the chemical reaction, and the crossover will occur by diffusion-controlled sound attenuation.

Finally, by plotting the dimensionless sound attenuation coefficient $\alpha(\omega)/\omega^2$ versus $\omega\tau$ at various temperatures, one can see whether $\tau_D \sim [(T - T_C)/T_C]^{-3\nu}$ or $\tau_C \sim ((T - T_C)/T_C]^{-\gamma}$ yields a better dynamic scaling function. The calculated change in sound velocity $\Delta C/C\omega^2$ for $\omega\tau << 1$ is proportional to $(\omega\tau)^{1/2}$ in the nonreactive case and to $\omega\tau$ in the reactive case [195] .

6.3 Light scattering from reactive systems

Light scattering as a probe of chemical dynamics is able to detect spontaneous fluctuations which are drastically increased near the critical points. Like sound, this is essentially a non-perturbative method which, by using the microscopic polarizability of molecules, provides a complimentary tool to other methods of studying chemical kinetics. Measuring the angular dependence and the shift in the frequency of the light scattered by a system allows one to find both the spatial and the time dependence of the fluctuations. The electromagnetic wave impinges upon a molecule inducing the dipole moment proportional to the polarizability. For a macroscopic continuum, the summation over the polarizabilities over individual molecules gives the macroscopic dielectric constant $\varepsilon = \varepsilon_0 + \delta\varepsilon\,(r,t)$, where the latter part comes from the space-time fluctuations of local thermodynamic variables. Assuming local equilibrium of a reactive binary mixture described by the reduced entropy $S_1 \equiv S - (\partial S/\partial\xi)_{T,p}\,\xi$, pressure p, and the extent of reaction ξ, one obtains

$$\delta\varepsilon\,(r,t) = \left(\frac{\partial\varepsilon}{\partial S}\right)_{p,\xi}\delta S_1 + \left(\frac{\partial\varepsilon}{\partial p}\right)_{S_1,\xi}\delta p + \left(\frac{\partial\varepsilon}{\partial\xi}\right)_{p,S_1}\delta\xi. \tag{6.34}$$

Light scattering theory assumes [196], [197] that the intensity I of scattered light having wavenumber K_0 at point R_0 is described by the expression

$$I\,(r_0,K,\omega) = \frac{NI_0K_0^4}{16\pi^2R_0^2}\sin^2\phi\,\langle\delta\varepsilon\,(K,\omega)\,\delta\varepsilon\,(-K,0)\rangle \tag{6.35}$$

where I_0 is the intensity of the incident light, N is the number of molecules in the scattered volume, ω is the shift in the frequency of the scattered light, K is the change in the wavenumber in the medium of the scattered light, and ϕ is the angle between the electric vector of the incident wave and the vector R_0. Inserting the space Fourier transform of Eq. (6.34) into (6.35) gives the light intensity as a function of time-dependent correlators of the statistical independent thermodynamic variables $S_1\,(K,t)$, $p\,(K,t)$ and $\xi\,(K,t)$. The hydrodynamic equations express the latter in terms of the static thermodynamic fluctuations of $S_1\,(K)\,,\,p\,(K)$ and $\xi\,(K)$.

There are two major mechanisms for the relaxation of fluctuations [197]: one arises from transport properties, and the other is due to the local changes induced by chemical reactions. The first mechanism depends upon spatial gradients, i.e., the line width of the scattered light is proportional to the square of the wavenumber K^2, whereas for the second mechanism, the line width is independent on K. Therefore, in principle, one can measure chemical relaxation times by the extrapolation $K \to 0$. However, in a real fluid, the kinetic and chemical modes are coupled. This coupling can be described by the hydrodynamic equations (6.22)–(6.25), which, according to (6.35), have to be written in the $\omega - K$ Fourier form,

$$\frac{d}{dt}\begin{pmatrix} \beta_1(K,t) \\ \beta_2(K,t) \\ \beta_3(K,t) \\ \beta_4(K,t) \end{pmatrix} = M \begin{pmatrix} \beta_1(K,t) \\ \beta_2(K,t) \\ \beta_3(K,t) \\ \beta_4(K,t) \end{pmatrix} \tag{6.36}$$

where the independent variables $\beta_i(K,t)$ are $div(v)$ and the normalized forms of deviations of pressure, reduced entropy, and the mass fraction of the species with the highest partial specific enthalpy from their equilibrium values [198],

$$\begin{aligned}
\beta_1(k,t) &= \left(\frac{\rho_0}{k_B v C_{p,\xi}}\right)^{1/2} S_1(k,t), \\
\beta_2(k,t) &= \left[\frac{\rho_0 (\partial A/\partial \xi)_{T.p}}{k_B T v}\right]^{1/2} \xi(K,t), \\
\beta_3(k,t) &= \left[\frac{(\partial v/\partial p)_{S,\xi}}{k_B T v}\right]^{1/2} p(k,t), \\
\beta_4(k,t) &= \left(\frac{\rho_0}{k_B T K^2 v}\right)^{1/2} div(v)
\end{aligned} \tag{6.37}$$

and the 4×4 matrix M has the following form

$$\begin{aligned}
M_{11} &= \frac{\lambda}{\rho_0 C_{p,\xi}} K^2 + \gamma_T \tau_{T,p}^{-1}, \\
M_{13} &= M_{31} = C_1 K^2 + C_2 \tau_{T,p}^{-1}, \\
M_{12} &= M_{21} = k_T D \left[\frac{(\partial A/\partial \xi)_{T.p}}{T C_{p,\xi}}\right]^{1/2} K^2 - \gamma_T^{1/2} \tau_{T,p}^{-1}, \\
M_{22} &= D K^2 + \tau_{T,p}^{-1}, \\
M_{23} &= M_{32} = C_3 K^2 + C_4 \tau_{T,p}^{-1}, \\
M_{33} &= C_5 K^2 + \left(r_M C_{p,A}/C_{p,\xi}\right) \tau_{T,p}^{-1}, \\
M_{34} &= M_{43} = K c_\infty, \\
M_{44} &= C_6 K^2, \\
M_{14} &= M_{24} = M_{41} = M_{42} = 0
\end{aligned} \tag{6.38}$$

where all notations can be found in [198]. However, when the damping terms in the matrix M are much smaller than the frequency of the sound wave Kc_∞,

$$\frac{\lambda}{\rho_0 c_{p,\xi}} K^2 << Kc_\infty,$$
$$DK^2 << Kc_\infty, \tag{6.39}$$
$$\frac{1}{\rho_0}\left(\frac{4}{3}\eta + \varsigma\right) K^2 << Kc_\infty$$

and the inverse chemical reaction time $\tau_{T,p}^{-1}$ satisfies

$$\tau_{T,p}^{-1} << Kc_\infty, \tag{6.40}$$

then the dispersion relation reduces in zero approximation to Eqs. (6.26)–(6.27) with additional first-order coupling between the pressure fluctuations and fluctuations in entropy and concentration. The four roots of the dispersion equation have been found [199]. The two real roots $z_{1,2}$ are related to the non-propagating modes and define the Rayleigh line in the scattered light,

$$z_{1,2} = \frac{1}{2}\left\{(d_1 + d_2) \pm \left[(d_1 - d_2)^2 + 4d_{12}^2\right]^{1/2}\right\}, \tag{6.41}$$

and the complex roots $z_{3,4}$, relate to two propagating modes defining two symmetric Brillouin lines,

$$z_{3,4} = \left[\Gamma K^2 + \frac{1}{2}\frac{c_\infty^2 - c_0^2}{c_0^2}\tau_{T,p}^{-1}\right] \pm iKc_\infty. \tag{6.42}$$

The following notation has been introduced in Eq. (6.41),

$$d_1 = \frac{\lambda}{\rho_0 C_{p,\xi}} K^2 + \frac{C_{p,A} - C_{p,\xi}}{C_{p,\xi}\tau_{T,p}}; \quad d_2 = DK^2 + \tau_{T,p}^{-1},$$
$$d_{12} = \left[Dk_T K^2 - \frac{T}{\tau_{T,p}}\left(\frac{\partial \xi}{\partial T}\right)_{p,A}\right]\sqrt{\left(\frac{\partial A}{\partial \xi}\right)_{T,p}(TC_{p,\xi})^{-1}} \tag{6.43}$$

while Γ in Eq. (6.42) is the same combination of transport coefficients, which is the damping factor of the Brillouin lines for a two-component non-reactive system.

Let us check the restrictions arising from inequalities (6.39) and (6.40). Conditions (6.39) are always met for sufficiently small K. For typical magnitudes of the transport coefficient and the speed of the sound, the conditions given in Eq. (6.39) are satisfied even for the maximum scattering wavenumbers ($K_{\max} \approx 2 \times 10^5$ sm^{-1}) encountered in light scattering experiments. To satisfy condition (6.40), K must be larger than some minimal value which depends on $\tau_{T,p}$. To be in the region of wavenumbers used in light scattering experiments, the inverse of the chemical reaction time must be of order 10^9 s^{-1} or smaller.

Spectroscopic techniques permits us to obtain qualitative information about the local environment around solute molecules. Using pyrene as a spectroscopic probe, appreciable enhancement up to 2.5 times the bulk density was seen for supercritical C_2H_6,CHF_3 and CO around pyrene [200].

6.4 Inhomogeneous structure of near-critical reactive systems

The high compressibility of a dilute solution near the liquid-gas critical point gives rise to a local density increase, in which the density of solvent around a solute molecule is higher that the bulk density. The measurement of partial molar volumes of naphthalene in ethylene yields [201] values that are extremely large and negative, implying that the local environment near the solvent molecules is very different from the bulk properties of the fluid. This phenomenon, called "molecular charisma" [201], or "density augmentation" [200], or "fluid clustering" [202], has been studied intensively using spectroscopic measurements. The use of pyrene as a solute is specially convenient because it is a chromophore, fluorescent, and its photo-physics is well known. Extensive measurements, using steady-state and time-resolved fluorescence of dissolved pyrene in supercritical CO_2 [203] and critical water [204], clearly show state-dependent local density enhancements in dilute supercritical fluid solutions. The magnitude of this enhancement was analyzed in terms of the local CO_2 density, based on the Onsager reaction field theory [205]. A solute with a dipole moment μ, contained in a spherical cavity of radius a, and dissolved in a solvent with static dielectric constant ϵ and refractive index n, polarizes the solvent and produces a reaction field. This reaction field lowers the ground-state energy of the solvated solute relative to the gas phase value. This shift $\Delta\nu$ between vapor and solution phase is [206]

$$\Delta\nu = \frac{\mu_g^2 - \mu_c^2}{a^3}\frac{n^2-1}{2n^2+1} + \frac{2\mu_c\left(\mu_g-\mu_c\right)}{a^3}\left(\frac{\epsilon-1}{\epsilon-2} - \frac{n^2-1}{n^2+2}\right) \tag{6.44}$$

where μ_g and μ_c are the dipole moments of the ground state and excited states, respectively. However, experiment shows that a shift $\Delta\nu$ between vapor and solution is not linear in $\left(n^2-1\right)/\left(n^2+2\right)$, as predicted by Eq. (6.44). This demonstrates that the local fluid environment surrounding the ground state pyrene molecules is different from that predicted by Eq. (6.44), and exceeds the expected value. On the other hand, there is a relationship [207] between the polarizability term $\left(n^2-1\right)/\left(n^2+2\right)$ and the supercritical CO_2 density ρ_{loc},

$$\frac{n^2-1}{n^2+2} = 6.6\rho_{loc} + 1.25\rho_{loc}^2 - 264\rho_{loc}^3. \tag{6.45}$$

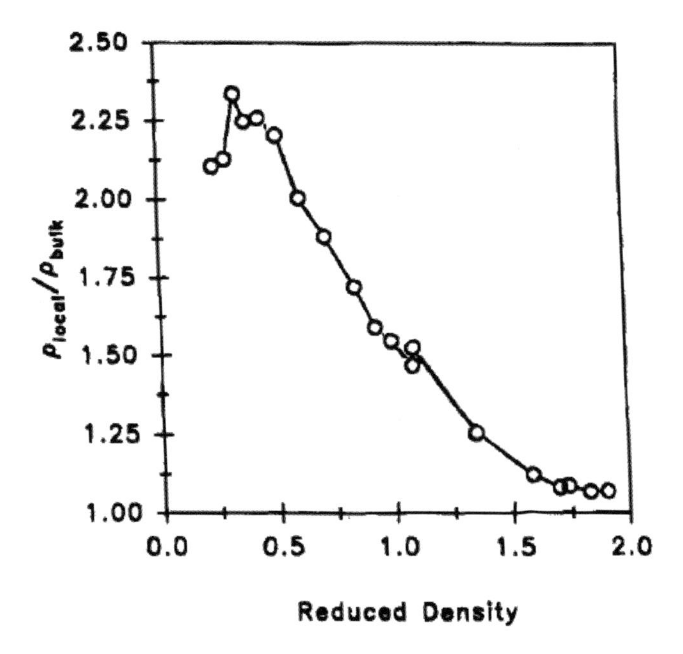

Fig. 6.1 Recovered density enhancement ρ_{loc}/ρ_{bulk} as a function of the dimensionless fluid density ρ/ρ_{crit} for pyrene in CO_2 at temperature $T = 45$ K. Reproduced from Ref. [203] with permission, copyright (1995), American Chemical Society.

Equations (6.44) and (6.45) were used [203] to find the spectral shift $\Delta\nu$ as a function of bulk density ρ_{bulk}. The experimental spectral shift $\Delta\nu$ is used to compute a local polarizability term in Eq. (6.44), which determines [203] the local density ρ_{loc} from Eq. (6.45). In this manner, the local density augmentation ρ_{loc}/ρ_{bulk} was found as a function of the reduced density ρ/ρ_{crit} of CO_2, as shown in Fig. 6.1, which displays a local density augmentation up to 170 at about half the critical density. In addition to this local density enhancement, connected with the ground state of pyrene molecules, similar experiments have been performed for the pyrene excited state. The latter turned out to be 1.5 times greater than in the ground state, due to the increased electrostatic interaction between the excited state of pyrene and CO_2. All these results do not depend on the pyrene concentration, indicating that pyrene molecules are solvated individually in CO_2.

The same increase of the solvent density around a solute molecule compared with the corresponded bulk density has been obtained [208] using (N,N-dimethylamino) benzonitrile as a probe instead of pyrene. Additional verification of the inhomogeneity of a near-critical reactive system has been obtained by indirect measurements combined with thermodynamic calculations [209], [210]. Finally, the same result follows from numerical simulations based on the Lennard-Jones potential [211] and the theory of integral equations [212].

Chapter 7

Conclusions

The main body of this book consists of a detailed discussion of the interconnection between "physical" critical phenomena and chemical reactions. Although these two phenomena appear to be quite different, deep analysis shows [22], [21] that the instability of different nonequilibrium phenomena (hydrodynamic and chemical instabilities, laser transitions, among others) show an intimate connection with equilibrium phase transitions and critical phenomena. Indeed, both are characterized by the enhancement of fluctuations, long-range order and critical slowing-down. Examples of this connection include the relation which has been established [213] between the laser threshold and the critical point of a second-order phase transition, and the analogy between phase transitions of the first and second order and hard and soft transitions of chemical reactions. The theory of critical phenomena, which is supported experimentally with a precision of few decimal places, dates from the 1960's and 1970's. It is not surprising that the methods used in this theory (renormalization group, mode-mode coupling, etc) have been successfully employed in modern chemistry.

Of three branches of modern physics — theory, experiment and simulation — we mostly used the first in the phenomenological analysis, with references to experimental work given in appropriate places. Many references to numerical methods are given in review articles, such as [165], which describe numerical methods of phase equilibria in reactive systems.

A near-critical fluid has some of the advantages of both a liquid and a gas: high density ensures high dissolving power which is accompanied by high diffusivity of solutes and low viscosity facilitates mass transport. In addition, high compressibility allows the tuning of the density and dissolving power by small changes in pressure. That is the reason why chemical reactions near the critical point are in such wide use in modern science and technology. Many applications can be found in review articles ([156], [165]). We may mention applications in chemistry (supercritical extraction [145]), metallurgy (the recovery of metals from coal ash [214], phase diagrams in alloy systems [215]), geochemistry (distribution and activity of species in aqueous solutions [216], solubility of minerals in brine solutions [217]), geology (equilibria between aqueous and mineral phases [218], mineral equilibria [219]), food industry (extraction of caffeine from coffee beans [220] and oil from corn, sunflower and peanuts [221]), and the pharmaceutical industry (extraction of vitamin E from soybean oil and its purification [222]).

This subject still attracts great interest, and it is very likely that many new applications will be found in the future.

Bibliography

[1] H. G. Harris and J. M. Praushnitz, Ind. Eng. Chem. Found. **8**, 180, 1969.
[2] M. Gitterman, Phys. Stat. Sol. **59**, 295, 1980.
[3] I. Prigogine and R. Defay, *Chemical Thermodynamics,* Longmans Green, New York, 1954
[4] L. D. Landau and E. M. Lifshitz, *Statistical Physics*, Pergamon, London, 1958; *Fluid Dynamics*, Pergamon, London, 1987.
[5] A. A. Galkin and V. V. Lunin, Russian Chem. Rev. **74**, 21, 2005.
[6] S. N. V. K. Aki and M. A. Abraham, Envir. Prog. **17**, 246, 1998.
[7] S. N. V. K. Aki and M. A. Abraham, Chem. Eng. Sci. **54,** 3533, 1999.
[8] A. Kruse and H. Vogel, Chem. Eng. Technol. **31**, 23, 2008.
[9] A. M. N. da Ponte, J. Supercrit. Fluids **47**, 344, 2009.
[10] M. Gitterman, Rev. Mod. Phys. **50**, 85, 1978.
[11] G. Arcovito, C. Falocci, M. Riberty, and L. Mistura, Phys. Rev. Lett. **22**, 1040, 1969.
[12] H. E. Stanley, *Introduction to Phase Transitions and Critical Phenomena*, Oxford University Press, 1993.
[13] V. L. Ginzburg, Sov. Phys. - Solid State **2**, 1824, 1960
[14] D. Yu. Ivanov, Doklady Physics **47**, 267, 2002
[15] D. Yu. Ivanov, L. A. Makarevich, and O. N. Sokolova, JETP Letters **20**, 121, 1974.
[16] W. Wagner, N. Kurzeja, and B. Pieperbeck, Fluid Phase Equilibria **79**, 151, 1992.
[17] L. A. Makarevich, O. N. Sokolova, and A. M. Rozen, Sov. Phys. - JETP **40**, 305, 1974.
[18] M. Gitterman, J. Stat. Phys. **58**, 707, 1990.
[19] J. S. Rowlinson and F. I. Swinton, *Liquids and Liquid Mixtures*, Butterworth, London, 1982.
[20] G. M. Schneider, Adv. Chem. Phys. **17**, 1, 1970.
[21] A. Nitzan, Phys. Rev. A **17**, 1513, 1978.
[22] A. Nitzan, P. Ortoleva, J. Deutch, and J. Ross, J. Chem. Phys. **60**, 1056, 1974.
[23] I. Matheson, D. F. Walls, and C. W. Gardiner, J. Stat. Phys. **12**, 21, 1975.

[24] A. S. Mikhailov, Phys. Lett. A **73**, 143, 1979; Z. Phys. B **41**, 277, 1981.
[25] W. Horsthemke and R. Levever, *Noise-induced Transitions*, Springer, Berlin, 1984.
[26] D. G. Peck, A. J. Mehta, and K. P. Johnston, J. Phys. Chem. **93**, 4297, 1998.
[27] B. Lin and A. Akgerman, Ind. Eng. Chem. Res. **40**, 1113, 2001.
[28] H. Li, B. Han, J. Liu,.L. Gao, Z. Hou, T. Jiang, Z. Liu, X. Zhang, and J. He, Chemistry- Eur. J. **8,** 5593, 2002.
[29] J. Quin, M. T. Timko, A. J. Allen, C. J. Russel, B. Winnik, B. Buckley, J. I. Steinfeld, and J. W. Tester, J. Am. Chem. Soc. **126**, 5465, 2004.
[30] B. Wang, B. Han, T. Jiang, Z. Zhang, Y. Xie, W. Li, and W. Wu, J. Phys. Chem. B **109**, 24203, 2005.
[31] T. Nishikawa, Y. Inoue, M. Sato, Y. Iwai, and Y. Arai, AIChE J. **44**, 1706, 1998.
[32] S. Glasstone, K. J. Laidler, and H. Eyring, *The Theory of Rate Processes*, 1st ed, McGraw-Hill, New York, 1941.
[33] C. A. Eckert, D. H. Ziger, K. P. Johnston, and S. J. Kim, J. Phys. Chem. **90**, 2738, 1986.
[34] C. Muller, A. Steiger, and F. Becker, Thermochim. Acta **151**, 131, 1989.
[35] L. Van Hove, Phys. Rev. **95**, 1374, 1954.
[36] I. Procaccia and M. Gitterman, Phys. Rev. Lett. **46**, 1163, 1981; Phys. Rev. A **25**, 1137, 1982.
[37] A. Z. Patashinskii, V. L. Pokrovskii, and M. V. Feigel'man, Sov. Phys. - JETP **55**, 851, 1982.
[38] R. G. Griffiths and J. C. Wheeler, Phys. Rev. A **2**, 1047, 1970.
[39] M. E. Paulaitis and G. C. Alexander, Pure Appl. Chem. **59**, 61, 1987.
[40] S. T. Milner and P. C. Martin, Phys. Rev. A **33**, 1996, 1986.
[41] A. Onuki and R. A. Ferrell, Physica A **164**, 245, 1989; B. Zappoli, D. Bailly, Y. Garrabos, R. Le Neindre, P. Guenoun, and D. Beysens, Phys. Rev. A **41**, 2264, 1990; H. I Boukari, J. N. Shaumeyer, M. E. Briggs, and R. W. Gammon, Phys. Rev. A **41**, 2260, 1990.
[42] R. P. Behringer, A. Onuki, and H. Meyer, J. Low Temp. Phys. **81**, 71, 1990; H. Meyer and F. Zhong, Compt. Rend. Mecanique **332**, 327, 2004.
[43] B. Zappoli and P. Carles, Acta Astron. **38**, 39, 1996.
[44] Yu. I. Dakhnovskii, A. A. Ovchinnikov, and V. A. Benderskii, Chem. Phys. **66**, 93, 1982.
[45] M. Sato, Y. Ikushima, K. Hatekada, and T. Ikeshoji, Int. J. Chem. Reactor Eng. **3,** A15, 2005.
[46] C. E. Bunker and Ya-Ping-Sun, J. Am. Chem. Soc. **117,** 10865, 1995.
[47] I. Raspo, S. Meradji, and B. Zappoli, Chem. Eng. Sci. **62**, 4182, 2007.
[48] S.-T. Chung, H.-C. Huang, S.-J. Pan, W.-T. Tsai, P.-Y. Lee, C.-H. Yang, and M.-B. Wu, Corrosion Science **50**, 2614, 2008.
[49] M. Gitterman, Physica A **386**, 1, 2007.
[50] M. Gittermam, Physica A **388**, 1046, 2009.
[51] M. Green, J. Chem. Phys. **22**, 398, 1954.
[52] H. Mori, Progr. Theor. Phys. **34**, 399, 1965.

[53] R. Zwanzig, *Lecture in Theoretical Physics*, vol. 3, Interscience, New York, 1961.
[54] T. Yamamoto, J. Chem. Phys. **33**, 281, 1960.
[55] B. J. Berne and R. Pecora, *Dynamic Light Scattering*, Wiley, NewYork, 1976.
[56] R. M. Velasco and L. S. Garcia-Colin, Physica **72**, 233, 1973.
[57] D. L. Carle, W. G. Laidlaw, and H. N. W. Lekkerkerker, Farad. Soc. Trans. **71**, 1448, 1975.
[58] R. D. Mountain and J. M. Deutch, J. Chem. Phys. **50**, 1103, 1969.
[59] L. Letamendia, J. P. Vindoula, C. Vaucamps, and G. Nouchi, Phys. Rev. A **41**, 3178, 1990.
[60] C. Allain and P. Lallemand, J. de Chimie Physique **77,** 881, 1980.
[61] A. Munster, *Statistical Thermodynamics*, Springer, 1974.
[62] J. K. Baird and J. C. Clunie, J. Phys. Chem. **102**, 6498, 1998.
[63] Y. W. Kim and J. K. Baird, Int. J. Thermophys. **22**, 1449, 2001.
[64] C. D. Specker, J. M. Ellis, and J. K. Baird, Int. J. Thermophys. **28**, 846, 2007.
[65] Y. W. Kim and J. K. Baird, Int. J. Thermophys. **25**, 1025, 2004.
[66] I. Procaccia and M. Gitterman, Phys. Rev. A **27**, 555, 1983; ibid. **30**, 647, 1984.
[67] J. C. Wheeler and R. G. Petschek, Phys. Rev. A **28**, 2442, 1983.
[68] I. R. Krichevskii, Yu. V. Tsekhanskaya, and Z. A. Polyakova, Rus. J. Phys. Chem. **40**, 715, 1966; ibid. **43**, 1393, 1969.
[69] G. Morrison, Phys. Rev. A **30**, 644, 1984.
[70] M. Gitterman and I. Procaccia, J. Chem. Phys. **78**, 2648, 1983; AIChE J. **29**, 686, 1983.
[71] I. Procaccia and M. Gitterman, J. Chem. Phys. **78**, 5275, 1983.
[72] T. Toroumi, J. Sakai, T. Kawakami, D. Osawa, and M. Azuma, J. Soc. Ind. Jpn. **49**, 1, 1946.
[73] S. C. Greer, Phys. Rev. A **31,** 3240, 1985.
[74] J. L. Tveekrem, R. H. Cohn, and S. C. Greer, J. Chem. Phys. **86**, 3602, 1987.
[75] R. Narayan and M. J. Antal Jr, J. Am. Chem. Soc. **112**, 1927, 1990.
[76] R. B. Snyder and C. A. Eckert, AIChE J. **19**, 1126, 1973.
[77] E. S. Gould, *Mechanism and Structure of Organic Chemistry*, Holt, Rinehart, and Winston, New York, 1959.
[78] Y. W. Kim and J. K. Baird, J. Phys. Chem. A **107**, 8435, 2003; ibid. **109**, 4750, 2005.
[79] J. L. Tveekrem and D. T. Jacobs, Phys. Rev. A **27**, 2773, 1983.
[80] R. H. Cohn and D. T. Jacobs, J. Chem. Phys. **80**, 856, 1984.
[81] D. T. Jacobs, J. Chem. Phys. **91**, 560, 1989.
[82] J. Ke, B. Han, M. W. George, H. Yan, and M. Poliakoff, J. Am. Chem. Soc. **123**, 3661, 2001.
[83] B. Hu, R. D. Richey, and J. K. Baird, J. Chem. Eng. Data **54**, 1537, 2009.
[84] M. E. Fisher, Phys. Rev. **176**, 257, 1968.
[85] M. Gitterman and V. Steinberg, Phys. Rev. A **22**, 1287, 1980.

[86] M. A. Anisimov, A. V. Voronel, and E. E. Gorodetskii, Sov. Phys. - JETP **33**, 605, 1971.
[87] J. Thoen, R. Kindt, W. van Dael, M. Merabet, and T. K. Bose, Physica A **156**, 92, 1989.
[88] E. U. Frank, in *The Physics and Chemistry of Aqueous Ionic Solutions*, eds. M.-C. Bellissent-Fuel and G. W. Neilson, Reidel, Dordecht, 1987, p. 337.
[89] A. Stein and G. F. Allen, J. Chem. Phys. **59**, 6079, 1973.
[90] C. H. Shaw and W. I. Goldburg, J. Chem. Phys. **65**, 4906, 1976.
[91] E. M. Anderson and S. C. Greer, Phys. Rev. A **30**, 3129, 1984.
[92] D. Jasnow, W. I. Goldberg, and J. S. Semura, Phys. Rev. A **9**, 355, 1974.
[93] M. E. Fisher and J. S. Langer, Phys. Rev. Lett. **20**, 665, 1968.
[94] M. Gitterman, Phys. Rev. A **28**, 358, 1983.
[95] J. C. Wheeler, Phys. Rev. A **30**, 648, 1984.
[96] G. R. Anderson and J. C. Wheeler, J. Chem. Phys. **69**, 2082, 3403, 1978; ibid. **73**, 5778, 1980.
[97] J. L. Tveecom, S. C. Greer, and D. T. Jacobs, Macromolecules **21**, 147, 1988.
[98] E. Gulari, B. Chu, and D. Woermann, J. Chem. Phys. **73**, 2480, 1980; P. Gansen, T. Janssen, W. Schon, D. Woermann, and H. Schonert, Ber. Bunsenges. Phys. Chem. **84**, 1149, 1980.
[99] K. Ohbayashi and B. Chu, J. Chem. Phys. **68**, 5066, 1978.
[100] C. M. Knobler and R. L. Scott, J. Chem. Phys. **76**, 2606, 1982.
[101] P. Gansen and D. Woermann, J. Phys. Chem. **88**, 2655, 1984.
[102] K. Kawasaki, in *Phase Transitions and Critical Phenomena*, vol 5a, C. Domb and M. S. Green, eds., Academic, New York, 1976.
[103] P. C. Hohenberg and B. I. Halpern, Rev. Mod. Phys. **49**, 435, 1977.
[104] G. Wilson and J. Kogut, Phys. Rep. C **12,** 75, 1974.
[105] Ya. B. Zel'dovich, Zh. Fiz. Khim. **11**, 685, 1938 (in Russian).
[106] R. Aris, Arch. Ration. Mech. Anal. **19**, 81, 1965.
[107] R. Aris, Arch. Ration. Mech. Anal. **27**, 356, 1968.
[108] J. M. Powers and S. Paulucci, Am. J. Phys. **76**, 848, 2008.
[109] E. B. Starikov and B. Norden, J. Phys. Chem. B **111**, 14431, 2007.
[110] A. L. Cornish-Bowden, J. Biosciences **27**, 121, 2002.
[111] M. Gitterman and V. Steinberg, J. Chem. Phys. **69,** 2763, 1978; Phys. Rev. Lett. **35**, 1588, 1975.
[112] M. Gitterman and V. Steinberg, Phys. Rev. A **20**, 1236, 1979.
[113] W. Ebeling and R. Sandig, Ann. Phys. (Leipzig) **28**, 289, 1973.
[114] Y. Albeck and M. Gitterman, Phil. Mag. B **56**, 881, 1997.
[115] H. S. Caram and L. E. Scriven, Chem. Ing. Sci. **31**, 163, 1976.
[116] G. Othmer, Chem. Ing. Sci. **31**, 993, 1976.
[117] L. R. Corrales and J. Wheeler, J. Chem. Phys. **91**, 7097, 1989.
[118] V. Talanquer, J. Chem. Phys. **96**, 5408, 1992.
[119] D. Borgis and M. Moreau, J. Stat. Phys. **50**, 935, 1988
[120] Y. Rabin and M .Gitterman, Phys. Rev. A **29**, 1496, 1984.
[121] H. Reiss and M. Shugard, J. Chem. Phys. **65**, 5280, 1976.

[122] M. Gitterman, in *Nonequilibrium Statistical Mechanics*, ed. E. S. Hernandez, World Scientific, 1989, p. 103.
[123] Y. Albeck and M. Gitterman, Phys. Rev. Lett. **60**, 588, 1988.
[124] J W. Cahn and J. E. Hillard, J. Chem. Phys. **28**, 258, 1958.
[125] M. Motoyama, J. Phys. Soc. Japan **65**, 1894, 1996; M. Motoyama and T. Ohta, ibid. **66**, 2715, 1997.
[126] R. Lefever, D. Carati, and N. Hassani, Phys. Rev. Lett. **75**, 1674, 1995.
[127] S. C. Glotzer, D. Stauffer, and N. Jan, Phys. Rev. Lett. **75**, 1675, 1995.
[128] S. C. Glotzer, E. A. DiMarzio, and M. Muthukumar, Phys. Rev. Lett. **74**, 2034, 1995.
[129] Y. Huo, X. Jiang, H. Zhang, and Y. Yang, J. Chem. Phys. **118**, 9830, 2003.
[130] S. Puri and H. L. Frish, J. Phys. A **27**, 6027,1994.
[131] S. C. Glotzer and A. Conigliuo, Phys. Rev. E **50**, 4241, 1994.
[132] S. C. Glotzer, D. Stauffer, and N. Jan, Phys. Rev. Lett. **72**, 4109, 1994.
[133] J. J. Christiansen, K. Elder, and H. C. Fogelby, Phys. Rev. E **54**, R2212, 1996.
[134] S. Toxvaerd, Phys. Rev. E **53**, 3710, 1996.
[135] D. Carati and R. Lefever, Phys. Rev. E **56**, 3127, 1997.
[136] C. Tong and Y. Yang, J. Chem. Phys. **116,** 1519, 2002.
[137] B. Liu, C. Tong, and Y. Yang, J. Phys. Chem. B **105**, 10091, 2001.
[138] D. Q. He, S. Kwak, and E. B. Nauman, Macromol. Theory Simul. **5**, 801, 1996.
[139] N. F. Mott, Rev. Mod. Phys. **40**, 677, 1968.
[140] P. Nozieres, Physica B **117**, 16, 1983.
[141] L. A. Turkevich and M. H. Cohen, J. Phys. Chem. **88**, 3751, 1984.
[142] W. Hefner and F. Hensel, Phys. Rev. Lett. **48**, 1026, 1982.
[143] V. Steinberg, A. Voronel, D. Linsky, and U. Schindewolf, Phys. Rev. Lett. **45**, 1338, 1980.
[144] U. Schindewolf, J. Phys. Chem. **88**, 3820, 1984.
[145] S. Peter, Ber. Bunsenges. Phys. Chem. **88**, 875, 1984.
[146] M. D. Palmieri, J. Chem. Edu. **65**, A254, 1988; ibid. **66**, A141, 1989.
[147] M. A. McHugh and V. J. Krukonis, *Supercritical Fluid Extraction, Principles and Practice*, Butterworths, Boston, 1986.
[148] L. T. Taylor, *Supercritical Fluid Extraction*, Wiley, New York, 1996.
[149] G. A. M. Diepen and F. E. C. Scheffer, J. Am. Chem. Soc. **70**, 4085, 1948.
[150] M. Gitterman, Am. J. Phys. **56**, 1000, 1988.
[151] C. A. Van Gunst, F. E. C. Scheffer, and G. A. M. Diepen, J. Phys. Chem. **57,** 578, 1953.
[152] S. K. Ma, *Modern Theory of Critical Phenomena*, Benjamin, Reading, Mass, 1976.
[153] U. van Wasen and G. M. Schneider, J. Phys. Chem. **84**, 229, 1980.
[154] M. McHugh and M. E. Paulatis, J. Chem. Eng. Data, **25**, 326, 1980.
[155] B. C. Wu, M. T. Klein, and S. I. Sandler, Ind. Eng. Chem. Res. **30**, 822, 1991.
[156] R. Fernandez-Prini and M. L. Japes, Chem. Soc. Rev. **23**, 155, 1994.

[157] G. Debenedetti, J. W. Tom, X. Kwauk, and S. D. Yeo, Fluid Phase Equilibria **83**, 311, 1993.
[158] T. M. Doscher and M. El-Arabi, Oil Gas J. **80**, 144,1982.
[159] R. K. Helling, and J. W. Tester, Envir. Sci. Technol. **22,** 1319, 1988.
[160] H. H. Yang and C. A. Eckert, Ind. Eng. Chem. Res. **27**, 2009, 1988.
[161] N. A. Collins, P. G, Debenedetti, and S. Sundaresan, AIChE J. **34**, 1211, 1988.
[162] W. Eisenbach, P. J. Gottsh, N. Niemann, and K. Zosel, Fluid Phase Equilibria **10**, 315, 1983.
[163] R. P. De Fillipi, V. J. Krukonis, and M. Modell, Environmental Protection Agency Report No. EPA-600/2-80-054, 1980.
[164] C. R. Yonker, R. W. Wright, S. L Frye, and R. D. Smith in *Supercritical Fluids*, Squires and Paulaitis, eds, American Chemical Society Symp. Series 329: Washington, 1987, p. 172.
[165] A. E. Mather, Fluid Phase Equilibria **30**, 83, 1986.
[166] D. Barbosa and M. F. Doherty, Proc. R. Soc. Lond. A **413**, 443, 1987,
[167] Stortenbekar, Z. Physik Chem. **10**, 183, 1892.
[168] B. Widom, private communication.
[169] V. A. Rotenberg and G. M. Kuznetsov, Rus. J. Phys. Chem. **48**, 2723, 1974.
[170] I. R. Krichevskii, E. E. Sominskaja, and N. L. Nechitallo, Sov. Phys.-Dokl. **282**, 541, 1985.
[171] I. R. Krichevskii, E. E. Sominskaja, N. L. Nechitallo and G. N. Lukashova, Sov. Phys.-Dokl. **295**, 811, 1987.
[172] G. N. Lewis, and M. Rendall, *Thermodynamics*, McGraw-Hill, New York, 1961.
[173] A. A. Sobyanin, Sov. Phys. - Uspechi, **29**, 570, 1986.
[174] B. C. McEwan, J. Chem. Soc. **123**, 2284, 1923.
[175] C. S. Hudson, Z. Phys. Chem. **47**, 113, 1904.
[176] R. J. L. Andon and J. D. Cox, J. Chem. Soc. 4601, 1952.
[177] F. Jona and G. Shirane, *Ferroelectric Crystals*, Pergamon, Oxford, 1962.
[178] J. C. Wheeler and B. Widom, J. Chem. Phys. **52**, 5334, 1970.
[179] R. E. Goldstein, J. Chem. Phys. **79**, 4439, 1983.
[180] G. A. Larsen and C. M. Sorensen, Phys. Rev. Lett. **54**, 343, 1985.
[181] P. K. Khabibullaev and A. A. Saidov, *Phase Separation in Soft Matter Physics*, Springer-Verlag, 2003.
[182] A. Einstein, in *Physical Acoustics*, R. B. Lindsey, ed., Dowden, Hutchison and Ross, Strasbourg, 1974.
[183] H. Dunker, D. Woermann, and J. K. Bhattacharjee, Ber. Busenges. Phys. Chem. **87**, 591, 1983.
[184] R. A. Ferrell and J. K. Bhattacharjee, Phys. Rev. A **31**, 1788, 1985; J. K. Bhattacharjee and R. A. Ferrell, Phys. Rev. A **24**, 1643, 1981.
[185] P. S. LaPlace, Ann. Chim. Phys. **3**, 238, 1816.
[186] R. A. Ferrell, N. Menyhard, H. Schmidt, F. Schwabl, and P. Szepfalusy, Phys. Rev. Lett. **18**, 891, 1967.
[187] Y. Harada, Y. Suzuki, and Y. Ishida, J. Phys. Soc. Jpn. **48**, 703, 1980.
[188] W. Mayer, S. Hoffman, G. Meier, and I. Alig, Phys. Rev. E **55**, 3102, 1997.

[189] I. Iwanowski, S. Z. Mirzaev, and U. Kaatze, Phys. Rev. E **73**, 061508, 2000.
[190] R. Behrends and U. Kautze, Europhys. Lett. **65**, 221, 2004.
[191] G. Sanchez and G. W. Garland, J. Chem. Phys. **79**, 3100, 1983.
[192] R. Behrends, T. Telgmann, and U. Kaatze, J. Chem. Phys. **117**, 9828, 2002.
[193] R. Behrends, U. Kaatze, and M. Schach, J. Chem. Phys. **119**, 7957, 2003.
[194] L. S. Garcia-Colin and S. M. T. de la Selva, Physica **75**, 37, 1974.
[195] H. G. E. Hentschel and I. Procaccia, J. Chem. Phys. **76**, 666, 1982
[196] R. D. Mountain, Rev. Mod. Phys. **38**, 205, 1966.
[197] L. Blum and Z. W. Zalzburg, J. Chem. Phys. **48**, 2292, 1968.
[198] H. N. W. Lekkerkerker and W. G. Laidlaw, Phys. Rev. A **9**, 346, 1974.
[199] H. N. W. Lekkerkerker and W. G. Laidlaw, Phys. Chem. Liquids **3**, 225, 1972.
[200] J. F. Brennecke, D. L. Tomasko, J. Peshkin, and C. A. Eckert, Ind. Eng. Chem. Res. **29**, 1682, 1990.
[201] C. A. Eckert and C. A. Knutson, Fluid Phase Equilibria **83**, 93, 1993.
[202] S. Kim and K. P. Johnston, AIChE J. **33**, 1603, 1987.
[203] J. K. Rice, E. D. Niemeyer. R. A. Dunbar, and F. V. Bright, J. Am. Chem. Soc. **117**, 5832, 1995.
[204] E. D. Niemeyer. R. A. Dunbar, and F. V. Bright, Appl. Spectroscopy **51**, 1547, 1997.
[205] L. Onsager, J. Am. Chem. Soc. **58**, 1486, 1936.
[206] P. A. Geldorf, R. P. H. Rettschnick, and G. J. Hoytnick, Chem. Phys. Lett. **4**, 59, 1969.
[207] Y.-P. Sun, C. E. Bunker, and N. B. Hamilton, Chem. Phys. Lett. **210**, 111, 1993.
[208] O. Kajimoto, M. Futakami, T. Kobayashi, and K. Yamasaki, J. Phys. Chem. **92**, 1347, 1988.
[209] S. Kim and K. P. Johnston, Ind. Eng. Chem. Res. **26**, 1206, 1987.
[210] C. R. Yonker and R. D. Smith, J. Phys. Chem. **92**, 2374, 1988.
[211] I. B. Petsche and P. G. Debenedetti, J. Chem. Phys. **91**, 7075, 1989.
[212] D. Cochran and L. L. Lee, in *Supercritical Fluid Science and Technology*, eds. K. P. Johnston and J. M. L. Penninger, ACS Symposium, Ser. 406, Washington, 1989, Chapter 3.
[213] R. Graham and H. Haken, Z. Phys. **237**, 31, 1970.
[214] A. K. Mehrotra, P. R. Bishnoi, and W. V. Svrcek, Can. J. Chem. Eng. **57**, 225, 1979.
[215] J. P. Holloway and J. F. Stubbins, Phil. Mag. A **52**, 475, 1985.
[216] D. K. Hordstrom and J. L. Munoz, *Geochemical Thermodynamics*, (second ed.), Blackwell Scientific Publications, Boston, 1994.
[217] C. E. Harvier and J. H. Weare, Geochimica Cosmochimica Acta **44**, 981, 1980.
[218] H. C. Helgeson, Pure Appl. Chem. **57**, 31, 1985.
[219] W. F. Giggenbach, Geochimica Cosmochimica Acta **45**, 393, 1981.
[220] J. R. Williams, B. Liu, and A. A. Clifford, *Supercritical Fluid Methods and Protocols,* Humana Press, 2000.

[221] L. Phan, H. Brown, J. White, A. Hodgson, and P. C. Jessop, Green Chem. **11**, 53, 2009.
[222] G. W. Burton and M. G. Trauber, Ann. Rev. Nutr. **10**, 357, 1990.

Index